Longman Mathematical Texts

Integral equations

Longman Mathematical Texts

Edited by Alan Jeffrey and Iain Adamson

Longman Mathematical Texts

Integral equations

B. L. Moiseiwitsch

Department of Applied Mathematics and Theoretical Physics,
The Queen's University of Belfast

Longman
London and New York

Longman Group Limited London

Associated companies, branches and representatives throughout the world

Published in the United States of America by Longman Inc., New York

© Longman Group Limited 1977

First published 1977

Library of Congress Cataloging in Publication Data

Moiseiwitsch, Benjamin Lawrence.
  Integral equations.

  (Longman mathematical texts)
  1. Integral equations. 2. Hilbert space.
3. Linear operators. I. Title.
QA431.M57   515′.45   76-10282
ISBN 0-582-44288-5

Printed at The Pitman Press, Bath Ltd.

Preface

Many problems arising in mathematics, and in particular in applied mathematics or mathematical physics, can be formulated in two distinct but connected ways, namely as differential equations or as integral equations. In the former approach the boundary conditions have to be imposed externally, whereas in the case of integral equations the boundary conditions are incorporated within the formulation and this confers a valuable advantage to the latter method. Moreover the integral equation approach leads quite naturally to the solution of the problem as an infinite series, known as the Neumann expansion, in which the successive terms arise from the application of an iterative procedure. The proof of the convergence of this series under appropriate conditions presents an interesting exercise in elementary analysis.

Integral equations have been of considerable significance in the history of mathematics. Thus Laplace and Fourier transforms are examples of integral equations of the first kind, while another interesting example is Abel's integral equation which is associated with Huygens' tautochrone problem and has a singular kernel.

The Hilbert transform also possesses a singular kernel. It arises from a boundary value problem in a plane and enables the solution of the Hilbert type of singular integral equation to be derived. Integral transforms in general often provide a convenient method for finding the solution to the class of integral equations having kernels of the difference, or convolution, type.

This book is mainly concerned with linear integral equations although a brief discussion of a simple type of non-linear equation is given at the end of the first chapter. The theory of linear integral equations of the second kind was developed originally by Volterra and Fredholm. In its more general form the analysis is carried out for Lebesgue square integrable functions since they form a Hilbert space. In this volume I have attempted to avoid unnecessary complications wherever possible by proving results for square integrable functions without usually specifying the sense in which the integration is to be carried out. Thus the book can be followed, for the most part, without distinguishing between Riemann and Lebesgue integration and this, I hope, will make it suitable for a wider range of mathematics students.

I have devoted two chapters to Hilbert space and linear operators in Hilbert space, the integral occurring in linear integral equations being an example of a linear operator. Hilbert space is of fundamental importance in mathematical physics since it provides the foundation for an axiomatic formulation of quantum mechanics. For this reason I have thought it worthwhile to discuss, even if rather briefly, some more general situations in which we are concerned with elements or vectors in an abstract Hilbert space acted upon by completely continuous linear operators, an example of which is the linear integral operator with square integrable kernel.

The final chapter is concerned with the theory of Hilbert and Schmidt on Hermitian integral operators with square integrable kernels. In this theory the solution of linear integral equations of the second kind is expanded in terms of the characteristic functions and values of the kernel. This is a common procedure in theoretical physics although its mathematical justification is often disregarded. The book concludes with a short discussion of variational principles and methods.

The equations are numbered consecutively in each chapter. When referring to an equation in another chapter, the number of the chapter is inserted as a prefix but if the referenced equation is in the same chapter the prefix is omitted.

The mathematical knowledge required to work through this book, and to do the problems at the end of each chapter, is that which an undergraduate student should possess as a result of attending elementary courses in analysis, complex variable and linear algebra. Thus the book should be suitable for students in their final year of an honours mathematics or mathematical physics course. The problems have been chosen so as to illuminate the theory given in the main text. They are not too difficult and to gain full advantage from the book the student is strongly advised to tackle them.

Contents

5: Integral equations with singular kernels

6: Hilbert space

7: Linear operators in Hilbert space

8: The resolvent

9: Fredholm theory

10: Hilbert-Schmidt theory

Classification of integral equations

1.1 Historical introduction

The name *integral equation* for any equation involving the unknown function $\phi(x)$ under the integral sign was introduced by du Bois-Reymond in 1888. However, the early history of integral equations goes back a considerable time before that to Laplace who, in 1782, used the integral transform

$$f(x) = \int_0^\infty e^{-xs}\phi(s)\, ds \tag{1}$$

to solve linear difference equations and differential equations.

In connection with the use of trigonometric series for the solution of heat conduction problems, Fourier in 1822 found the reciprocal formulae

$$f(x) = \sqrt{\frac{2}{\pi}} \int_0^\infty \sin xs\, \phi(s)\, ds, \tag{2}$$

$$\phi(s) = \sqrt{\frac{2}{\pi}} \int_0^\infty \sin xs\, f(x)\, dx \tag{3}$$

and

$$f(x) = \sqrt{\frac{2}{\pi}} \int_0^\infty \cos xs\, \phi(s)\, ds, \tag{4}$$

$$\phi(s) = \sqrt{\frac{2}{\pi}} \int_0^\infty \cos xs\, f(x)\, dx \tag{5}$$

where the Fourier sine transform (3) and the cosine transform (5) provide the solutions $\phi(s)$ of the integral equations (2) and (4) respectively in terms of the known function $f(x)$.

In 1826 Abel solved the integral equation named after him having the form

$$f(x) = \int_a^x (x-s)^{-\alpha}\phi(s)\, ds \tag{6}$$

where $f(x)$ is a continuous function satisfying $f(a) = 0$, and $0 < \alpha < 1$.

For $\alpha = \frac{1}{2}$ Abel's integral equation corresponds to the famous tautochrone problem first solved by Huygens, namely the determination of the shape of the curve with a given end point along which a particle slides under gravity in an interval of time which is independent of the starting position on the curve. Huygens showed that this curve is a cycloid.

An integral equation of the type

$$\phi(x) = f(x) + \lambda \int_0^x k'(x-s)\phi(s)\,\mathrm{d}s \tag{7}$$

in which the unknown function $\phi(s)$ occurs outside as well as before the integral sign and the variable x appears as one of the limits of the integral, was obtained by Poisson in 1826 in a memoir on the theory of magnetism. He solved the integral equation by expanding $\phi(s)$ in powers of the parameter λ but without establishing the convergence of this series. Proof of the convergence of such a series was produced later by Liouville in 1837.

Dirichlet's problem, which is the determination of a function ψ having prescribed values over a certain boundary surface S and satisfying Laplace's equation $\nabla^2 \psi = 0$ within the region enclosed by S, was shown by Neumann in 1870 to be equivalent to the solution of an integral equation. He solved the integral equation by an expansion in powers of a certain parameter λ. This is similar to the procedure used earlier by Poisson and Liouville, and corresponds to a method of successive approximations.

In 1896 Volterra gave the first general treatment of the solution of the class of linear integral equations bearing his name and characterized by the variable x appearing as the upper limit of the integral.

A more general class of linear integral equations having the form

$$\phi(x) = f(x) + \int_a^b K(x, s)\phi(s)\,\mathrm{d}s \tag{8}$$

which includes Volterra's class of integral equations as the special case given by $K(x, s) = 0$ for $s > x$, was first discussed by Fredholm in 1900 and subsequently, in a classic article by him, in 1903. He employed a similar approach to that introduced by Volterra in 1884. In this method the Fredholm equation (8) is regarded as the limiting form as $n \to \infty$ of a set of n linear algebraic equations

$$\phi(x_r) = f(x_r) + \sum_{s=1}^n K(x_r, x_s)\phi(x_s)\delta_n \qquad (r = 1, \ldots, n) \tag{9}$$

where $\delta_n = (b-a)/n$ and $x_r = a + r\delta_n$. The solution of these equations can be readily obtained and Fredholm verified by direct substitution in the integral equation (8) that his limiting formula for $n \to \infty$ gave the true solution.

1.2 Linear integral equations

Integral equations which are linear involve the integral operator

$$L = \int_a^b K(x, s)\, ds \tag{10}$$

having the *kernel* $K(x, s)$. It satisfies the linearity condition

$$L[\lambda_1\phi_1(s) + \lambda_2\phi_2(s)] = \lambda_1 L[\phi_1(s)] + \lambda_2 L[\phi_2(s)] \tag{11}$$

where

$$L[\phi(s)] = \int_a^b K(x, s)\phi(s)\, ds \tag{12}$$

and λ_1 and λ_2 are constants.

Linear integral equations are named after Volterra and Fredholm as follows:

The *Fredholm equation of the first kind* has the form

$$f(x) = \int_a^b K(x, s)\phi(s)\, ds \qquad (a \le x \le b) \tag{13}$$

Examples are given by Laplace's integral (1) and the Fourier integrals (2) and (4).

The *Fredholm equation of the second kind* has the form

$$\phi(x) = f(x) + \int_a^b K(x, s)\phi(s)\, ds \qquad (a \le x \le b) \tag{14}$$

while its corresponding *homogeneous equation* is

$$\phi(x) = \int_a^b K(x, s)\phi(s)\, ds \qquad (a \le x \le b) \tag{15}$$

The *Volterra equation of the first kind* has the form

$$f(x) = \int_a^x K(x, s)\phi(s)\, ds \qquad (a \le x \le b) \tag{16}$$

An example of such an equation is Abel's equation (6) which, however, has a special feature arising from the presence of a *singular* kernel

$$K(x, s) = (x - s)^{-\alpha} \qquad (0 < \alpha < 1)$$

with a singularity at $s = x$.

The *Volterra equation of the second kind* has the form

$$\phi(x) = f(x) + \int_a^x K(x, s)\phi(s)\,\mathrm{d}s \qquad (a \leq x \leq b) \qquad (17)$$

We see that the Volterra equations can be obtained from the corresponding Fredholm equations by setting $K(x, s) = 0$ for

$$a \leq x < s \leq b.$$

It can be readily seen also that the Volterra equation (16) of the first kind can be transformed into one of the second kind by differentiation, for we have

$$f'(x) = K(x, x)\phi(x) + \int_a^x \frac{\partial}{\partial x} K(x, s)\phi(s)\,\mathrm{d}s$$

so that provided $K(x, x)$ does not vanish in $a \leq x \leq b$ we obtain

$$\phi(x) = \frac{f'(x)}{K(x, x)} - \int_a^x \left[\frac{\partial}{\partial x} K(x, s)/K(x, x)\right]\phi(s)\,\mathrm{d}s$$

1.3   Special types of kernel

1.3.1   Symmetric kernels

Integral equations with symmetric kernels satisfying

$$K(s, x) = K(x, s) \qquad (18)$$

possess certain simplifying features. For this reason it is valuable to know that the integral equation

$$\psi(x) = g(x) + \int_a^b P(x, s)\mu(s)\psi(s)\,\mathrm{d}s \qquad (19)$$

with the unsymmetrical kernel $P(x, s)\mu(s)$, where however $P(s, x) = P(x, s)$, can be transformed into the integral equation (14) with symmetric kernel

$$K(x, s) = \sqrt{\mu(x)}\, P(x, s)\sqrt{\mu(s)} \qquad (20)$$

by multiplying (19) across by $\sqrt{\mu(x)}$ and setting

$$\phi(x) = \sqrt{\mu(x)}\,\psi(x) \tag{21}$$

and

$$f(x) = \sqrt{\mu(x)}\,g(x). \tag{22}$$

Real symmetric kernels are members of a more general class known as Hermitian kernels which are not necessarily real and are characterized by the relation

$$\overline{K(s, x)} = K(x, s), \tag{23}$$

the bar denoting complex conjugate. We shall investigate the properties of integral equations with Hermitian kernels in chapter 10.

1.3.2 Kernels producing convolution integrals

A class of integral equation which is of particular interest has a kernel of the form

$$K(x, s) = k(x - s) \tag{24}$$

depending only on the difference between the two coordinates x and s. The type of integral which arises is

$$\int_{-\infty}^{\infty} k(x - s)\phi(s)\,\mathrm{d}s \tag{25}$$

called a convolution or folding. This nomenclature comes from the situation which occurs in Volterra equations where the integral (25) becomes

$$\int_{0}^{x} k(x - s)\phi(s)\,\mathrm{d}s \tag{26}$$

and the range of integration can be regarded as if it were folded at $s = x/2$ so that the point s corresponds to the point $x - s$ as shown in Fig. 1.

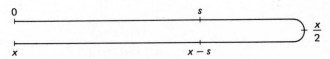

Fig. 1. Convolution or folding

Integral equations of the convolution type can be solved by using various kinds of integral transform such as the Laplace and Fourier transforms and will be discussed in detail in chapter 3.

1.3.3   Separable kernels

Another interesting type of integral equation has a kernel possessing the separable form

$$K(x, s) = \lambda u(x)\overline{v(s)} \tag{27}$$

The Fredholm integral equation of the second kind (14) now becomes

$$\phi(x) = f(x) + \lambda u(x)\int_a^b \overline{v(s)}\phi(s)\,\mathrm{d}s \tag{28}$$

and can be readily solved exactly. Thus we have, on multiplying (28) across by $\overline{v(x)}$ and integrating:

$$\int_a^b \overline{v(x)}\phi(x)\,\mathrm{d}x = \int_a^b \overline{v(x)}f(x)\,\mathrm{d}x + \lambda\int_a^b \overline{v(x)}u(x)\,\mathrm{d}x\int_a^b \overline{v(s)}\phi(s)\,\mathrm{d}s$$

which gives

$$\int_a^b \overline{v(x)}\phi(x)\,\mathrm{d}x = \frac{\int_a^b \overline{v(x)}f(x)\,\mathrm{d}x}{1 - \lambda\int_a^b \overline{v(x)}u(x)\,\mathrm{d}x} \tag{29}$$

so that

$$\phi(x) = f(x) + \frac{\lambda u(x)\int_a^b \overline{v(s)}f(s)\,\mathrm{d}s}{1 - \lambda\int_a^b \overline{v(t)}u(t)\,\mathrm{d}t} \tag{30}$$

We see that the solution (30) can be expressed in the form

$$\phi(x) = f(x) + \lambda\int_a^b R(x, s; \lambda)f(s)\,\mathrm{d}s \tag{31}$$

where

$$R(x, s; \lambda) = \frac{u(x)\overline{v(s)}}{1 - \lambda\int_a^b \overline{v(t)}u(t)\,\mathrm{d}t} \tag{32}$$

and is called the solving kernel or *resolvent* kernel.

The homogeneous equation (15) corresponding to (14) becomes in the present case

$$\phi(x) = \lambda u(x) \int_a^b \overline{v(s)} \phi(s) \, ds. \tag{33}$$

The solution $\phi(x)$ of equation (33) must satisfy

$$\int_a^b \overline{v(x)} \phi(x) \, dx = \lambda \int_a^b \overline{v(x)} u(x) \, dx \int_a^b \overline{v(s)} \phi(s) \, ds$$

The values of λ for which the homogeneous equation has solutions are called *characteristic values*. There exists just one value of λ for which (33) possesses a solution. This characteristic value λ_1 is given by

$$1 = \lambda_1 \int_a^b \overline{v(x)} u(x) \, dx, \tag{34}$$

the associated characteristic solution of (33) being $\phi_1(x) = cu(x)$ where c is an arbitrary constant.

The simple separable form (27) is a special case of the class of degenerate kernels

$$K(x, s) = \lambda \sum_{i=1}^n u_i(x) \overline{v_i(s)} \tag{35}$$

giving rise to integral equations which can be solved in closed analytical form as we shall show in chapter 9.

Example 1. As a simple illustration of an integral equation with separable kernel (27) we consider

$$\phi(x) = 1 + \lambda \int_0^1 xs\phi(s) \, ds \tag{36}$$

Here $f(x) = 1$, $K(x, s) = \lambda xs$ and so $u(x) = x$ and $v(s) = s$. It follows that the resolvent kernel is

$$R(x, s; \lambda) = \frac{xs}{1 - \lambda \int_0^1 t^2 \, dt}$$

$$= \frac{xs}{1 - \lambda/3} \tag{37}$$

and hence the solution to (36) is

$$\phi(x) = 1 + \frac{\lambda x}{1 - \lambda/3} \int_0^1 s \, ds$$

$$= 1 + \frac{3\lambda x}{2(3 - \lambda)} \qquad (\lambda \neq 3) \qquad (38)$$

Example 2.   Another example of an integral equation with a separable kernel is

$$\phi(x) = e^x + \lambda \int_0^1 e^{i\alpha(x-s)} \phi(s) \, ds \qquad (39)$$

where $f(x) = e^x$, $K(x, s) = \lambda e^{i\alpha(x-s)}$ so that $u(x) = e^{i\alpha x}$ and $v(s) = e^{i\alpha s}$. Then the resolvent kernel is

$$R(x, s; \lambda) = \frac{e^{i\alpha(x-s)}}{1 - \lambda} \qquad (40)$$

and hence the solution of (39) is

$$\phi(x) = e^x + \frac{\lambda e^{i\alpha x}}{1 - \lambda} \int_0^1 e^{(1-i\alpha)s} \, ds$$

$$= e^x + \frac{\lambda e^{i\alpha x}(e^{1-i\alpha} - 1)}{(1 - \lambda)(1 - i\alpha)} \qquad (41)$$

1.4   Square integrable functions and kernels

Functions $\phi(x)$ which are square integrable in the interval $a \leq x \leq b$ satisfy the condition

$$\int_a^b |\phi(x)|^2 \, dx < \infty \qquad (42)$$

where the integral is taken to be Riemann, or Lebesgue for greater generality. In the former case it is said that $\phi(x)$ is an R^2 function and in the latter case that it is an L^2 function. Continuous functions are square integrable over a finite interval since they are bounded. However the converse is not necessarily true, that is square integrable functions need not be continuous or bounded.

Kernels $K(x, s)$ defined in $a \leq x \leq b, a \leq s \leq b$ are said to be square integrable if they satisfy

$$\int_a^b \int_a^b |K(x, s)|^2 \, dx \, ds < \infty \qquad (43)$$

together with

$$\int_a^b |K(x, s)|^2 \, ds < \infty \qquad (a \leqslant x \leqslant b) \tag{44}$$

and

$$\int_a^b |K(x, s)|^2 \, dx < \infty \qquad (a \leqslant s \leqslant b) \tag{45}$$

where the integrals are taken to be Riemann or Lebesgue. The kernel is then called R^2 or L^2 respectively.

Kernels of particular interest are those which are *singular*. Thus consider a Volterra kernel of the form

$$K(x, s) = \begin{cases} \dfrac{F(x, s)}{(x - s)^\alpha} & (x > s) \\ 0 & (x < s) \end{cases} \tag{46}$$

where $F(x, s)$ is a continuous function and $0 < \alpha < 1$. Then $|F(x, s)| \leqslant M$ where M is a constant and so

$$\int_a^b \int_a^b |K(x, s)|^2 \, dx \, ds = \int_a^b dx \int_a^x ds \frac{|F(x, s)|^2}{(x - s)^{2\alpha}}$$

$$\leqslant M^2 \int_a^b dx \int_a^x ds (x - s)^{-2\alpha}$$

$$= \frac{M^2}{1 - 2\alpha} \int_a^b dx (x - a)^{1 - 2\alpha} \qquad (\alpha < \tfrac{1}{2})$$

$$= \frac{M^2 (b - a)^{2 - 2\alpha}}{2(1 - 2\alpha)(1 - \alpha)} \tag{47}$$

and so the double integral is finite if $0 < \alpha < \tfrac{1}{2}$. However if $\alpha \geqslant \tfrac{1}{2}$ the singular kernel (46) is not square integrable.

1.5   Singular integral equations

An integral equation of the type

$$\phi(x) = f(x) + \lambda \int_a^b K(x, s) \phi(s) \, ds \tag{48}$$

is said to be singular if the range of definition is infinite e.g., $0 < x < \infty$ or $-\infty < x < \infty$, or if the kernel is not square integrable.

Non-singular equations have a discrete *spectrum*, that is the associated homogeneous equation

$$\phi(x) = \lambda \int_a^b K(x, s)\phi(s)\, ds \qquad (49)$$

has non-trivial solutions $\phi_\nu(x)$ for a finite or at most a countable, although infinite, set of characteristic values λ_ν of the parameter λ. Also each characteristic value λ_ν has a finite *rank* (or index), that is it has a finite number of linearly independent characteristic functions $\phi_\nu^{(1)}(x), \ldots, \phi_\nu^{(p)}(x)$.

However if the integral equation is singular by virtue of having an infinite range of definition, the spectrum of values of λ may include a continuous segment. For example it may be readily verified that the homogeneous equation

$$\phi(x) = \lambda \int_{-\infty}^\infty e^{-|x-s|}\phi(s)\, ds \qquad (50)$$

has solutions of the form

$$\phi(x) = c_1 e^{-\sqrt{1-2\lambda}\, x} + c_2 e^{\sqrt{1-2\lambda}\, x} \qquad (51)$$

for the continuous spectrum of values $0 < \lambda < \infty$. Equation (50) is of the convolution or difference type and will be discussed further in section 3.2 of the chapter on integral equations of the convolution type.

Also if the integral equation has an infinite range of definition the characteristic values may have an *infinite rank*. Thus consider the equation

$$\phi(x) = \lambda \sqrt{\frac{2}{\pi}} \int_0^\infty \cos xs\, \phi(s)\, ds \qquad (52)$$

with kernel

$$K(x, s) = \sqrt{\frac{2}{\pi}} \cos xs \qquad (53)$$

Using the Fourier cosine transform (5) and its reciprocal formula (4) we have

$$\lambda\phi(x) = \sqrt{\frac{2}{\pi}} \int_0^\infty \cos xs\, \phi(s)\, ds$$

and so

$$\phi(x) = \lambda^2 \phi(x)$$

giving $\lambda^2 = 1$, or $\lambda = \pm 1$ as characteristic values. Their ranks are infinite since it can be shown without difficulty that

$$\phi_{\pm}(x) = \sqrt{\frac{\pi}{2}} e^{-ax} \pm \frac{a}{a^2 + x^2} \quad (x > 0) \tag{54}$$

are characteristic functions corresponding to the characteristic values $\lambda = \pm 1$ for all values of $a > 0$. The kernel (53) belongs to a class of singular kernels known as *Weyl* kernels.

1.6   Non-linear equations

In the case of non-linear equations, the spectrum of characteristic values λ may have the interesting property that the number of solutions of the integral equation changes as we pass through particular values of λ known as *bifurcation points*.

As an example we consider the simple non-linear equation

$$\phi(x) - \lambda \int_0^1 s\{\phi(s)\}^2 \, ds = 1 \tag{55}$$

and investigate its real solutions $\phi(x)$. We put

$$\int_0^1 s\{\phi(s)\}^2 \, ds = \alpha \tag{56}$$

in which case we have

$$\phi(x) = 1 + \lambda\alpha \tag{57}$$

so that

$$\alpha = (1 + \lambda\alpha)^2 \int_0^1 s \, ds = \tfrac{1}{2}(1 + \lambda\alpha)^2$$

This yields

$$\lambda^2\alpha^2 + 2(\lambda - 1)\alpha + 1 = 0 \tag{58}$$

which gives

$$\alpha = \frac{1 - \lambda \pm \sqrt{(\lambda - 1)^2 - \lambda^2}}{\lambda^2} \tag{59}$$

Hence

$$\phi(x) = \frac{1 \pm \sqrt{1 - 2\lambda}}{\lambda} \tag{60}$$

Thus there are two real solutions of the non-linear integral equation (55) if $\lambda < \frac{1}{2}$, one real solution $\phi(x) = 2$ if $\lambda = \frac{1}{2}$ and no real solutions if $\lambda > \frac{1}{2}$. This means that $\lambda = \frac{1}{2}$ is a bifurcation point. When $\lambda = 0$ one of the solutions is $\phi(x) = 1$ while the other solution is infinite. Thus $\lambda = 0$ is a *singular point*.

We now look at the associated homogeneous equation

$$\phi(x) = \lambda \int_0^1 s\{\phi(s)\}^2 \, ds \tag{61}$$

This gives

$$\phi(x) = \lambda \alpha \tag{62}$$

so that

$$1 = \lambda^2 \alpha \int_0^1 s \, ds = \tfrac{1}{2}\lambda^2 \alpha$$

yielding

$$\alpha = \frac{2}{\lambda^2} \tag{63}$$

and the solution

$$\phi(x) = \frac{2}{\lambda}. \tag{64}$$

In general non-linear equations present considerable difficulties and we shall not consider them again after this chapter.

Problems

1. Transform the Volterra integral equation of the first kind

$$x = \int_0^x (e^x + e^s)\phi(s) \, ds$$

into a Volterra equation of the second kind. Show that the solution ϕ satisfies a first order differential equation and hence solve the integral equation.

2. Solve the Fredholm equation of the second kind

$$\phi(x) = x + \lambda \int_0^1 \phi(s) \, ds$$

and show that the characteristic value of the associated homogeneous equation is $\lambda_1 = 1$.

3. Solve the Fredholm equation

$$\phi(x) = 1 + \lambda \int_0^1 e^{x+s} \phi(s) \, ds$$

and show that the characteristic value of the associated homogeneous equation is $\lambda_1 = 2/(e^2 - 1)$.

4. Solve the Fredholm equation

$$\phi(x) = 1 + \lambda \int_0^1 x^n s^m \phi(s) \, ds$$

and show that the characteristic value of the associated homogeneous equation is $\lambda_1 = m + n + 1$.

5. Solve the Fredholm equation

$$\phi(x) = x + \lambda \int_0^\pi \sin nx \sin ns \, \phi(s) \, ds$$

where n is an integer, and show that the characteristic value of the associated homogeneous equation is $\lambda_1 = 2/\pi$.

6. Show that the singular integral equation

$$\phi(x) = \lambda \sqrt{\frac{2}{\pi}} \int_0^\infty \sin xs \, \phi(s) \, ds$$

has characteristic solutions

$$\phi_\pm(x) = \sqrt{\frac{\pi}{2}} e^{-ax} \pm \frac{x}{a^2 + x^2} \quad (x > 0)$$

associated with characteristic values $\lambda = \pm 1$ for all $a > 0$.

7. Solve the non-linear equation

$$\phi(x) = 1 + \lambda \int_0^1 \{\phi(s)\}^2 \, ds$$

and show that $\lambda = \frac{1}{4}$ is a bifurcation point while $\lambda = 0$ is a singular point of the spectrum of characteristic values.

Connection with differential equations

2.1 Linear differential equations

We first observe that the *first order* differential equation

$$\frac{\mathrm{d}\phi}{\mathrm{d}x} = F(x, \phi) \qquad (a \leqslant x \leqslant b) \tag{1}$$

can be written immediately as the Volterra integral equation of the second kind

$$\phi(x) = \phi(a) + \int_a^x F[s, \phi(s)] \, \mathrm{d}s. \tag{2}$$

As an interesting but simple example of the above we consider the following problem, solved by Johannes Bernoulli for the $n = 3$ case:

To find the equation $y = \phi(x)$ of the curve joining a fixed origin O to a point P such that the area under the curve is $1/n$th of the area of the rectangle OXPY having OP as diagonal for all points P on the curve (see Fig. 2).

Evidently this problem is equivalent to solving the homogeneous Volterra integral equation of the second kind

$$\int_0^x \phi(s) \, \mathrm{d}s = \frac{1}{n} x\phi(x) \tag{3}$$

for the unknown function $\phi(x)$ where $\phi(0) = 0$.
This equation can be solved by converting it into the differential equation

$$\phi(x) = \frac{1}{n} \{\phi(x) + x\phi'(x)\}$$

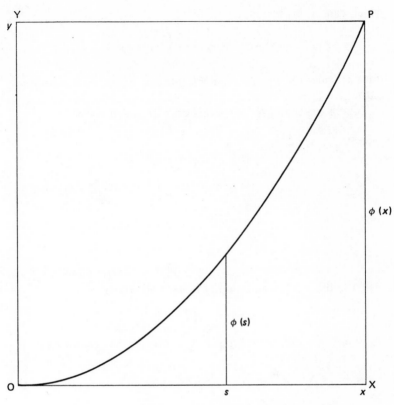

Fig. 2. Bernoulli's problem for $n = 3$; $y = x^2$.

or more simply

$$\phi'(x) = \frac{n-1}{x} \phi(x) \qquad (4)$$

which has the solution

$$\phi(x) = Ax^{n-1} \qquad (5)$$

For the case $n = 3$ solved by Bernoulli the curve is a parabola $y = Ax^2$.

We next consider the *second order differential equation*

$$\frac{d^2\phi}{dx^2} = F(x, \phi) \qquad (a \le x \le b) \qquad (6)$$

which can be expressed as

$$\phi(x) = \phi(a) + (x - a)\phi'(a) + \int_a^x (x - s)F[s, \phi(s)]\,ds, \tag{7}$$

this particular form being obtained on carrying out an integration by parts.

Taking $a = 0$ and $b - a = l$ this may be rewritten as

$$\phi(x) = \phi(0) + \frac{\phi(l) - \phi(0)}{l}\,x$$

$$+ \int_0^x (x - s)F[s, \phi(s)]\,ds + \frac{x}{l}\int_0^l (s - l)F[s, \phi(s)]\,ds \tag{8}$$

since

$$l\phi'(0) = \phi(l) - \phi(0) + \int_0^l (s - l)F[s, \phi(s)]\,ds.$$

If the differential equation is linear and we write

$$F[x, \phi(x)] = r(x) - q(x)\phi(x) \tag{9}$$

we obtain the Fredholm integral equation of the second kind

$$\phi(x) = f(x) + \int_0^l K(x, s)q(s)\phi(s)\,ds \tag{10}$$

where

$$f(x) = \phi(0) + \frac{\phi(l) - \phi(0)}{l}\,x - \int_0^l K(x, s)r(s)\,ds \tag{11}$$

and

$$K(x, s) = \begin{cases} \dfrac{s(l - x)}{l} & (s \leqslant x) \\[2ex] \dfrac{x(l - s)}{l} & (s \geqslant x) \end{cases} \tag{12}$$

In the particular case of a *flexible string* having mass $\rho(x)$ per unit length, stretched at tension T and vibrating with angular frequency ω, we have $q(x) = \omega^2\rho(x)/T$ and $r(x) = 0$. If the string has fixed ends at $x = 0$ and l, the transverse displacement $\phi(x)$ satisfies $\phi(0) = \phi(l) = 0$ so that $f(x) = 0$. Thus we arrive at the homogeneous Fredholm integral equation

$$\phi(x) = \int_0^l K(x, s)q(s)\phi(s)\,\mathrm{d}s \tag{13}$$

An important distinction between the differential equation and the equivalent integral equation approach should be observed. In the case of the differential equation formulation of a physical problem the boundary conditions are imposed separately whereas the integral equation formulation contains the boundary conditions implicitly. Thus, for example, we see at once from (13) and (12) that $\phi(x)$ vanishes at $x = 0$ and $x = l$.

Let us now consider the second order linear differential equation

$$\boldsymbol{L}v(x) = r(x) \qquad (0 \leqslant x \leqslant l) \tag{14}$$

where $\boldsymbol{L}$ is the linear differential operator given by

$$\boldsymbol{L} = \frac{\mathrm{d}}{\mathrm{d}x}\left\{ p(x)\frac{\mathrm{d}}{\mathrm{d}x} \right\} + q(x) \tag{15}$$

and $p(x)$ has no zeros in the range of definition $0 \leqslant x \leqslant l$.

We shall treat this equation somewhat differently from the previous equations by setting

$$\phi(x) = \frac{\mathrm{d}^2 v}{\mathrm{d}x^2}. \tag{16}$$

Then we have

$$\frac{\mathrm{d}v}{\mathrm{d}x} = v'(0) + \int_0^x \phi(s)\,\mathrm{d}s \tag{17}$$

and

$$v(x) = v(0) + xv'(0) + \int_0^x (x - s)\phi(s)\,\mathrm{d}s, \tag{18}$$

from which it follows that the differential equation may be expressed as the Volterra integral equation of the second kind

$$\phi(x) = f(x) + \int_0^x K(x, s)\phi(s)\,\mathrm{d}s \tag{19}$$

where

$$f(x) = \{p(x)\}^{-1}[r(x) - v'(0)p'(x) - \{v(0) + xv'(0)\}q(x)] \tag{20}$$

and

$$K(x, s) = \{p(x)\}^{-1}[(s-x)q(x) - p'(x)]. \tag{21}$$

2.2   Green's function

We consider the second order linear differential equation

$$\boldsymbol{L}v(x) = r(x) \qquad (a \leqslant x \leqslant b), \tag{22}$$

where $\boldsymbol{L}$ is the linear differential operator (15), and suppose that its solution $v(x)$ satisfies homogeneous boundary conditions at $x = a$ and $x = b$, namely

$$\alpha v(a) + \beta v'(a) = 0$$
$$\alpha v(b) + \beta v'(b) = 0 \tag{23}$$

where α and β are prescribed constants.

Let

$$\boldsymbol{L}v_i(x) = 0 \qquad (i = 1, 2) \tag{24}$$

where $v_1(x)$ and $v_2(x)$ are two linearly independent functions which satisfy the boundary conditions (23) prescribed at $x = a$ and b respectively. We shall show that the solution of (22) may be expressed in the form

$$v(x) = \int_a^b G(x, s)r(s) \, \mathrm{d}s \tag{25}$$

where

$$G(x, s) = \begin{cases} \dfrac{1}{A} v_1(x)v_2(s) & (x \leqslant s) \\[2mm] \dfrac{1}{A} v_2(x)v_1(s) & (x \geqslant s) \end{cases} \tag{26}$$

and A is a certain constant. $G(x, s)$ is called the Green's function.

The method we shall use involves the Dirac delta function $\delta(x)$ which vanishes for $x \neq 0$ and satisfies

$$\int_a^b \delta(x - s)r(s) \, \mathrm{d}s = r(x) \tag{27}$$

Clearly $\delta(x)$ is not a function as defined in the usual sense but belongs to a class known as *generalized functions*. A useful representation of the Dirac function is

$$\delta(x) = \lim_{h \to 0} I(h, x) \tag{28}$$

where $I(h, x)$ is the impulse function defined as

$$I(h, x) = \begin{cases} \dfrac{1}{h} & (0 \leqslant x \leqslant h) \\ 0 & \text{otherwise.} \end{cases} \tag{29}$$

Thus the representation (28) vanishes for $x \neq 0$ and has infinite magnitude at $x = 0$. Further

$$\lim_{h \to 0} \int_{-\infty}^{\infty} I(h, x) \, \mathrm{d}x = 1, \tag{30}$$

since the integral is unity for all values of h, and we write this as

$$\int_{-\infty}^{\infty} \delta(x) \, \mathrm{d}x = 1 \tag{31}$$

although care has to be exercised when interchanging the order of the limiting operation and the integration.

Now (25) has to provide a solution of (22) and so

$$\boldsymbol{L}G(x, s) = \delta(x - s). \tag{32}$$

It follows that

$$\lim_{\varepsilon \to 0} \int_{s-\varepsilon}^{s+\varepsilon} \boldsymbol{L}G(x, s) \, \mathrm{d}x = 1 \tag{33}$$

which leads to

$$\lim_{\varepsilon \to 0} \left[\frac{\partial}{\partial x} G(x, s) \right]_{x=s-\varepsilon}^{x=s+\varepsilon} = \frac{1}{p(s)}, \tag{34}$$

that is,

$$v_1(s)v_2'(s) - v_2(s)v_1'(s) = \frac{A}{p(s)} \tag{35}$$

or

$$\begin{vmatrix} v_1(s) & v_2(s) \\ v_1'(s) & v_2'(s) \end{vmatrix} = \frac{A}{p(s)} \tag{36}$$

where the determinant on the left-hand side is called the *Wronskian*. For A to be non-vanishing it is evident that v_1 and v_2 must be linearly independent.

As a simple example we consider the operator

$$\boldsymbol{L} = -c\frac{\mathrm{d}^2}{\mathrm{d}x^2} \qquad (0 \leqslant x \leqslant l) \tag{37}$$

and the end conditions $v(0) = v(l) = 0$. Then $p(x) = -c$ and $v_1(x) = x$, $v_2(x) = l - x$ so that $A = cl$ giving for the Green's function

$$G(x, s) = \begin{cases} \dfrac{x(l-s)}{cl} & (x \leqslant s) \\[2ex] \dfrac{(l-x)s}{cl} & (x \geqslant s) \end{cases} \tag{38}$$

2.3 Influence function

We consider a string of length l at tension T. The displacement $G(x, s)$ of the string at the point with coordinate x, resulting from the application of a unit force perpendicular to the string at the point with coordinate s, is called the influence function.

Now the displacement $G(s, s)$ of the string at the point s, assumed to be small, satisfies the equilibrium equation (see Fig. 3):

$$TG(s, s)\left(\frac{1}{s} + \frac{1}{l-s}\right) = 1 \tag{39}$$

which yields

$$G(s, s) = \frac{s(l-s)}{Tl} \tag{40}$$

and so it follows that the influence function is given by

$$G(x, s) = \begin{cases} \dfrac{x(l-s)}{Tl} & (x \leqslant s) \\[2ex] \dfrac{s(l-x)}{Tl} & (x \geqslant s) \end{cases} \tag{41}$$

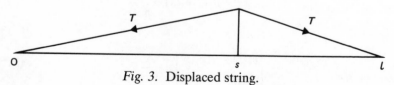

Fig. 3. Displaced string.

Comparison with (38) shows that the influence function derived above is just the Green's function for the linear differential operator

$$\boldsymbol{L} = -T\frac{\mathrm{d}^2}{\mathrm{d}x^2} \tag{42}$$

Now suppose that we apply a loading force $F(x)$ per unit length to the string. Then the displacement of the string at the point with coordinate x is given by

$$\phi(x) = \int_0^l G(x, s)F(s)\,\mathrm{d}s \tag{43}$$

using the principle of superposition. In the case of a horizontal wire at tension T, possessing mass $\rho(x)$ per unit length, the loading due to the force of gravity is $F(x) = -g\rho(x)$ and so the displacement of the wire becomes

$$\phi(x) = -g\int_0^l G(x, s)\rho(s)\,\mathrm{d}s. \tag{44}$$

As a further example we consider a *shaft* of length l and mass $\rho(x)$ per unit length. If the loading at the point with coordinate s is given by $F(s)$ per unit length and $G(x, s)$ is the influence function of the shaft, then the displacement $\phi(x)$ at the point x is again given by (43). Let us now suppose that the shaft is rotating with angular velocity ω. Then the loading due to the rotation is given by the centrifugal force

$$F(s) = \omega^2\rho(s)\phi(s), \tag{45}$$

and so we obtain the following homogeneous integral equation for the displacement of the shaft

$$\phi(x) = \omega^2\int_0^l G(x, s)\rho(s)\phi(s)\,\mathrm{d}s \tag{46}$$

The balancing of the elastic and centrifugal forces can occur only for certain discrete values of ω known as the *critical speeds* which correspond to values of ω^2 for which the integral equation (46) has non-vanishing solutions, that is its characteristic values.

Problems

1. Solve the Volterra equations of the second kind

$$\text{(i)} \quad \phi(x) = 1 + \int_0^x \phi(s)\, ds$$

$$\text{(ii)} \quad \phi(x) = 1 + 2\int_0^x s\phi(s)\, ds$$

by finding the solutions of the equivalent differential equations.

2. Solve the Volterra equation

$$x\phi(x) = x^n + \int_0^x \phi(s)\, ds$$

3. Solve the Volterra equation

$$\phi(x) = f(x) + \lambda \int_0^x (x - s)\phi(s)\, ds$$

for (i) $f(x) = 1$, (ii) $f(x) = x$ and $\lambda = \pm 1$.

4. Solve the Volterra equation

$$\phi(x) = f(x) + \lambda \int_0^x \sin(x - s)\phi(s)\, ds$$

for (i) $f(x) = x$ and $\lambda = 1$, (ii) $f(x) = e^{-x}$ and $\lambda = 2$.

5. Show that the Volterra equation of the first kind

$$f(x) = \int_0^x \frac{(x - s)^{n-1}}{(n-1)!}\, \phi(s)\, ds$$

has the continuous solution $\phi(s) = f^{(n)}(s)$ when $f(x)$ has continuous derivatives $f'(x)$, $f''(x)$, ..., $f^{(n)}(x)$, and $f(x)$ and its first $n - 1$ derivatives vanish at $x = 0$.

6. Show that the nth order linear differential equation

$$\frac{d^n y}{dx^n} - \lambda \sum_{r=1}^n a_r(x)\frac{d^{n-r}y}{dx^{n-r}} = f(x)$$

where $y(0) = y'(0) = \cdots = y^{(n-1)}(0) = 0$, is equivalent to the Volterra equation of the second kind

$$\phi(x) = f(x) + \lambda \int_0^x \sum_{r=1}^n \frac{a_r(x)(x-s)^{r-1}}{(r-1)!} \phi(s) \, ds$$

where $\phi(x) = d^n y/dx^n$.

7. Show that the solution of the differential equation

$$\frac{d^2 \phi}{dx^2} + \omega^2 \phi = F[x, \phi(x)] \qquad (0 \leqslant x \leqslant l)$$

obeying the end conditions $\phi(0) = \phi(l) = 0$, satisfies the integral equation

$$\phi(x) = \int_0^l G(x, s) F[s, \phi(s)] \, ds$$

where

$$G(x, s) = \begin{cases} -(\omega \sin \omega l)^{-1} \sin \omega x \sin \omega (l-s) & (x \leqslant s) \\ -(\omega \sin \omega l)^{-1} \sin \omega (l-x) \sin \omega s & (x \geqslant s) \end{cases}$$

is the Green's function for the linear differential operator $d^2/dx^2 + \omega^2$.

3

Integral equations of the convolution type

In the present chapter we shall be concerned with integral equations whose kernels have the form

$$K(x, s) = k(x - s) \tag{1}$$

which is a function only of the difference between the two coordinates x and s. The method of solution generally involves the use of integral transforms. Hence before proceeding to consider the various kinds of convolution type integral equations we shall briefly discuss the more important forms of integral transform.

3.1 Integral transforms

We have already mentioned Laplace and Fourier transforms in the historical introduction given in chapter 1.

The *Laplace transform* of a function $\phi(x)$ is given by

$$\Phi(u) = \int_0^\infty e^{-ux} \phi(x) \, dx \tag{2}$$

and the basic result which enables us to solve integral equations of the convolution type is the *convolution theorem* which states that if $K(u)$ is the Laplace transform of $k(x)$ then $K(u)\Phi(u)$ is the Laplace transform of

$$\int_0^x k(x - s)\phi(s) \, ds \tag{3}$$

To verify that this is correct, without any attempt at a rigorous argument, we see that the Laplace transform of (3) is

$$\int_0^\infty e^{-ux} \, dx \int_0^x k(x - s)\phi(s) \, ds = \int_0^\infty e^{-ut} k(t) \, dt \int_0^\infty e^{-us} \phi(s) \, ds$$

$$= K(u)\Phi(u)$$

on setting $t = x - s$ and noting that we may take $k(x)$ to vanish for $x < 0$.

In addition to the sine and cosine transforms (2) and (4) referred to in section 1.1, we have the exponential *Fourier transform* of a function $\phi(x)$ defined as

$$\Phi(u) = \frac{1}{\sqrt{2\pi}} \int_{-\infty}^{\infty} e^{iux} \phi(x) \, dx \tag{4}$$

and the reciprocal formula

$$\phi(x) = \frac{1}{\sqrt{2\pi}} \int_{-\infty}^{\infty} e^{-iux} \Phi(u) \, du. \tag{5}$$

Then, if $K(u)$ is the Fourier transform of $k(x)$, the convolution theorem for Fourier transforms states that $\sqrt{2\pi}\, K(u)\Phi(u)$ is the Fourier transform of

$$\int_{-\infty}^{\infty} k(x-s)\phi(s) \, ds \tag{6}$$

which can be easily verified as for the Laplace transform.

Another valuable integral transform is the *Mellin transform*

$$\Phi(u) = \int_{0}^{\infty} x^{u-1} \phi(x) \, dx \tag{7}$$

and the corresponding convolution theorem states that $K(u)\Phi(1-u)$ is the Mellin transform of

$$\int_{0}^{\infty} k(xs)\phi(s) \, ds \tag{8}$$

where $K(u)$ is the Mellin transform of $k(x)$. In fact, again without any attempt at rigour, we have that the Mellin transform of (8) is

$$\int_{0}^{\infty} x^{u-1} \, dx \int_{0}^{\infty} k(xs)\phi(s) \, ds = \int_{0}^{\infty} t^{u-1} k(t) \, dt \int_{0}^{\infty} s^{-u}\phi(s) \, ds$$

$$= K(u)\Phi(1-u)$$

on setting $t = xs$.

Throughout the following sections we shall assume that all the functions arising in the integral equations satisfy suitable conditions which permit all the transformations to be performed with validity.*

---

* For details see E. C. Titchmarsh, *Introduction to the Theory of Fourier Integrals*, 2nd ed., Oxford University Press, 1948.

3.2 Fredholm equation of the second kind

We consider firstly Fredholm equations of the type

$$\phi(x) = f(x) + \int_{-\infty}^{\infty} k(x-s)\phi(s)\, ds \qquad (-\infty < x < \infty) \qquad (9)$$

A formal solution of this equation can be obtained by introducing the Fourier transforms

$$\Phi(u) = \frac{1}{\sqrt{2\pi}} \int_{-\infty}^{\infty} e^{iux}\phi(x)\, dx, \qquad (10)$$

$$F(u) = \frac{1}{\sqrt{2\pi}} \int_{-\infty}^{\infty} e^{iux}f(x)\, dx, \qquad (11)$$

$$K(u) = \frac{1}{\sqrt{2\pi}} \int_{-\infty}^{\infty} e^{iux}k(x)\, dx. \qquad (12)$$

Using the convolution theorem

$$K(u)\Phi(u) = T\frac{1}{\sqrt{2\pi}} \int_{-\infty}^{\infty} k(x-s)\phi(s)\, ds \qquad (13)$$

where T denotes the Fourier integral operator

$$T = \frac{1}{\sqrt{2\pi}} \int_{-\infty}^{\infty} e^{iux}\, dx, \qquad (14)$$

we obtain

$$\Phi(u) = F(u) + \sqrt{2\pi}\, K(u)\Phi(u) \qquad (15)$$

as a result of operating with T on both sides of (9). This gives

$$\Phi(u) = \frac{F(u)}{1 - \sqrt{2\pi}\, K(u)} \qquad (16)$$

which provides the solution to the integral equation (9) in the form

$$\phi(x) = T^{-1}[\Phi(u)]$$
$$= \frac{1}{\sqrt{2\pi}} \int_{-\infty}^{\infty} \frac{e^{-ixu}F(u)}{1 - \sqrt{2\pi}\, K(u)}\, du \qquad (17)$$

on using the reciprocal formula (5).

Let us now consider the homogeneous equation

$$\phi(x) = \int_{-\infty}^{\infty} k(x-s)\phi(s)\,ds. \tag{18}$$

The solution to this can be written

$$\phi(x) = \sum_{\nu} \sum_{p=1}^{n} c_{\nu,p} x^{p-1} e^{-i\omega_\nu x} \tag{19}$$

where the $c_{\nu,p}$ are constants, the ω_ν are the zeros of $1 - \sqrt{2\pi}\,K(u)$ and n is the order of the multiplicity of the zero ω_ν.

Thus we have that

$$1 = \int_{-\infty}^{\infty} k(t) e^{i\omega_\nu t}\,dt \tag{20}$$

on setting $u = \omega_\nu$ in (12), and

$$0 = \int_{-\infty}^{\infty} k(t) t^{p-1} e^{i\omega_\nu t}\,dt \qquad (p = 2, \ldots, n) \tag{21}$$

on differentiating both sides of (12) $p-1$ times with respect to u and setting $u = \omega_\nu$. Hence

$$1 = \int_{-\infty}^{\infty} k(t)\left(1 - \frac{t}{x}\right)^{p-1} e^{i\omega_\nu t}\,dt \tag{22}$$

and so, setting $t = x - s$, we obtain

$$x^{p-1} e^{-i\omega_\nu x} = \int_{-\infty}^{\infty} k(x-s) s^{p-1} e^{-i\omega_\nu s}\,ds \tag{23}$$

which shows that each term in the sum on the right-hand side of (19) is a solution of (18).

Example 1. As a first example we consider the case

$$f(x) = e^{-|x|} \tag{24}$$

$$k(x) = \begin{cases} \lambda e^x & (x < 0) \\ 0 & (x > 0) \end{cases} \tag{25}$$

Then

$$F(u) = \frac{1}{\sqrt{2\pi}} \int_{-\infty}^{\infty} e^{-|x|+iux}\,dx$$

$$= \sqrt{\frac{2}{\pi}} \frac{1}{1+u^2} \tag{26}$$

while

$$K(u) = \frac{\lambda}{\sqrt{2\pi}} \int_{-\infty}^{0} e^{x+iux} \, dx$$

$$= \frac{\lambda}{\sqrt{2\pi}} \frac{1}{1+iu} \tag{27}$$

and so it follows that the solution is given by

$$\phi(x) = \frac{1}{\pi} \int_{-\infty}^{\infty} \frac{e^{-ixu} \, du}{(1-iu)(1-\lambda+iu)} \tag{28}$$

Then if $0 < \lambda < 1$ we may apply Cauchy's residue theorem to evaluate this integral obtaining the particular solution

$$\phi(x) = \begin{cases} \dfrac{2e^{-x}}{2-\lambda} & (x > 0) \\[2mm] \dfrac{2e^{(1-\lambda)x}}{2-\lambda} & (x < 0) \end{cases} \tag{29}$$

Since there is just one zero, having order of multiplicity 1, of $1 - \sqrt{2\pi}\, K(u) = (1-\lambda+iu)/(1+iu)$ at $u = i(1-\lambda)$, it follows that the solution of the associated homogeneous equation

$$\phi(x) = \lambda \int_{x}^{\infty} e^{x-s} \phi(s) \, ds \tag{30}$$

is just

$$\phi(x) = Ce^{(1-\lambda)x} \tag{31}$$

as can be readily verified by inspection.

Hence the general solution is given by

$$\phi(x) = \begin{cases} \dfrac{2e^{-x}}{2-\lambda} + Ce^{(1-\lambda)x} & (x > 0) \\[3mm] \left(\dfrac{2}{2-\lambda} + C\right)e^{(1-\lambda)x} & (x < 0) \end{cases} \tag{32}$$

where λ is confined to the range of values $0 < \lambda < 1$.

Example 2. As a second example we discuss the *Lalesco – Picard* equation for which

$$k(x) = \lambda e^{-|x|}. \tag{33}$$

We shall suppose that $f(x)$ and hence $F(u)$ are unknown, in which case it is convenient to rewrite the solution (17) in the form

$$\phi(x) = f(x) + \int_{-\infty}^{\infty} e^{-ixu} F(u) M(u) \, du \tag{34}$$

where

$$M(u) = \frac{K(u)}{1 - \sqrt{2\pi} \, K(u)} \tag{35}$$

and

$$f(x) = \frac{1}{\sqrt{2\pi}} \int_{-\infty}^{\infty} e^{-ixu} F(u) \, du. \tag{36}$$

For then we have, using the convolution theorem for Fourier transforms

$$\int_{-\infty}^{\infty} e^{-ixu} F(u) M(u) \, du = \int_{-\infty}^{\infty} f(u) m(x - u) \, du \tag{37}$$

where

$$m(x) = \frac{1}{\sqrt{2\pi}} \int_{-\infty}^{\infty} e^{-ixu} M(u) \, du, \tag{38}$$

that the solution (34) is given by

$$\phi(x) = f(x) + \int_{-\infty}^{\infty} f(u) m(x - u) \, du. \tag{39}$$

Now in the present example

$$K(u) = \frac{1}{\sqrt{2\pi}} \int_{-\infty}^{\infty} \lambda e^{-|x| + iux} \, dx$$

$$= \sqrt{\frac{2}{\pi}} \frac{\lambda}{1 + u^2}, \tag{40}$$

and so

$$M(u) = \sqrt{\frac{2}{\pi}} \frac{\lambda}{1 + u^2 - 2\lambda} \tag{41}$$

from which it follows that

$$m(x) = \frac{1}{\pi} \int_{-\infty}^{\infty} \frac{\lambda e^{-ixu}}{1+u^2-2\lambda} \, du$$

$$= \frac{\lambda}{\sqrt{1-2\lambda}} \, e^{-\sqrt{1-2\lambda}\,|x|} \tag{42}$$

provided $\lambda < \frac{1}{2}$ and using Cauchy's residue theorem. Hence a particular solution of the integral equation is

$$\phi(x) = f(x) + \frac{\lambda}{\sqrt{1-2\lambda}} \int_{-\infty}^{\infty} f(u) e^{-\sqrt{1-2\lambda}\,|x-u|} \, du. \tag{43}$$

Turning our attention to the homogeneous equation

$$\phi(x) = \lambda \int_{-\infty}^{\infty} e^{-|x-s|} \phi(s) \, ds \tag{44}$$

we note that there are two zeros, both having multiplicity of order 1, of

$$1 - \sqrt{2\pi}\, K(u) = \frac{1+u^2-2\lambda}{1+u^2} \tag{45}$$

at $u = \pm i\sqrt{1-2\lambda}$, so that the solution of (44) takes the form

$$\phi(x) = c_1 e^{-\sqrt{1-2\lambda}\,x} + c_2 e^{\sqrt{1-2\lambda}\,x} \tag{46}$$

where, for the integral occurring in (44) to have a meaning, the real part of $\sqrt{1-2\lambda}$ must satisfy

$$\operatorname{Re}\sqrt{1-2\lambda} < 1. \tag{47}$$

Thus all values of λ in the range $0 < \lambda < \infty$ are allowed, but if $\lambda > \frac{1}{2}$ the solution may be more suitably written as

$$\phi(x) = a_1 \sin(\sqrt{2\lambda-1}\,x) + a_2 \cos(\sqrt{2\lambda-1}\,x). \tag{48}$$

However when $\lambda = \frac{1}{2}$, $1 - \sqrt{2\pi}\, K(u)$ has a single zero at $u = 0$ with multiplicity of order 2 in which case the general solution of the homogeneous equation is

$$\phi(x) = c_1 + c_2 x. \tag{49}$$

3.3 Volterra equation of the second kind

If we set $f(x) = 0$, $k(x) = 0$, and $\phi(x) = 0$ for $x < 0$ in the Fredholm equation (9) of the second kind considered in section 3.2 we obtain the integral equation

$$\phi(x) = f(x) + \int_0^x k(x - s)\phi(s)\,\mathrm{d}s \qquad (x > 0) \qquad (50)$$

which is a Volterra equation of the second kind with a convolution type integral.

Equation (50) can be solved most conveniently by introducing the Laplace transforms

$$\Phi(u) = \int_0^\infty e^{-ux}\phi(x)\,\mathrm{d}x, \qquad (51)$$

$$F(u) = \int_0^\infty e^{-ux}f(x)\,\mathrm{d}x, \qquad (52)$$

$$K(u) = \int_0^\infty e^{-ux}k(x)\,\mathrm{d}x. \qquad (53)$$

Applying the convolution theorem for Laplace transforms

$$K(u)\Phi(u) = \boldsymbol{L}\int_0^x k(x - s)\phi(s)\,\mathrm{d}s \qquad (54)$$

where $\boldsymbol{L}$ denotes the Laplace integral operator

$$\boldsymbol{L} = \int_0^\infty e^{-ux}\,\mathrm{d}x, \qquad (55)$$

we find that

$$\Phi(u) = F(u) + K(u)\Phi(u) \qquad (56)$$

which yields

$$\begin{aligned}
\Phi(u) &= \frac{F(u)}{1 - K(u)} \\
&= F(u) + F(u)M(u) \qquad (57)
\end{aligned}$$

where

$$M(u) = \frac{K(u)}{1 - K(u)}. \qquad (58)$$

Hence the solution of (50) can be written as

$$\phi(x) = f(x) + \int_0^x m(x-s)f(s)\,ds \tag{59}$$

where

$$M(u) = \int_0^\infty e^{-ux} m(x)\,dx. \tag{60}$$

Example 1. We examine first the simple case where the kernel is

$$k(x) = \begin{cases} \lambda x & (x > 0) \\ 0 & (x < 0) \end{cases} \tag{61}$$

Then we have

$$K(u) = \lambda \int_0^\infty e^{-ux} x\,dx = \frac{\lambda}{u^2} \tag{62}$$

so that

$$\begin{aligned} M(u) &= \frac{\lambda}{u^2 - \lambda} \\ &= \frac{\sqrt{\lambda}}{2}\left(\frac{1}{u - \sqrt{\lambda}} - \frac{1}{u + \sqrt{\lambda}}\right) \end{aligned} \tag{63}$$

and hence

$$m(x) = \frac{\sqrt{\lambda}}{2}(e^{\sqrt{\lambda}x} - e^{-\sqrt{\lambda}x}). \tag{64}$$

Thus the solution to the integral equation (50), as given by (59), is

$$\phi(x) = f(x) + \frac{\sqrt{\lambda}}{2}\int_0^x f(s)\{e^{\sqrt{\lambda}(x-s)} - e^{-\sqrt{\lambda}(x-s)}\}\,ds \tag{65}$$

Example 2. Our next example has the kernel

$$k(x) = \begin{cases} \lambda e^x & (x > 0) \\ 0 & (x < 0) \end{cases} \tag{66}$$

Then

$$K(u) = \lambda \int_0^\infty e^{(1-u)x}\,dx = \frac{\lambda}{u-1} \tag{67}$$

so that

$$M(u) = \frac{\lambda}{u - (\lambda + 1)} \qquad (68)$$

and hence

$$m(x) = \lambda e^{(\lambda+1)x}. \qquad (69)$$

Thus, using (59) the solution of the integral equation (50) is

$$\phi(x) = f(x) + \lambda \int_0^x f(s) e^{(\lambda+1)(x-s)} \, ds \qquad (70)$$

Example 3. Our last example has the kernel

$$k(x) = \begin{cases} \lambda \sin x & (x > 0) \\ 0 & (x < 0) \end{cases} \qquad (71)$$

Then

$$K(u) = \lambda \int_0^\infty e^{-ux} \sin x \, dx = \frac{\lambda}{u^2 + 1} \qquad (72)$$

and so

$$M(u) = \frac{\lambda}{u^2 + 1 - \lambda} \qquad (73)$$

which yields

$$m(x) = \begin{cases} \dfrac{\lambda}{\sqrt{1-\lambda}} \sin(\sqrt{1-\lambda}\, x) & (\lambda < 1) \\[2ex] \lambda x & (\lambda = 1) \\[2ex] \dfrac{\lambda}{\sqrt{\lambda-1}} \sinh(\sqrt{\lambda-1}\, x) & (\lambda > 1) \end{cases} \qquad (74)$$

on using the appropriate Laplace transform formulae. Hence the solution of the integral equation (50) is given by

$$\phi(x) - f(x) = \begin{cases} \dfrac{\lambda}{\sqrt{1-\lambda}} \displaystyle\int_0^x f(s) \sin\{\sqrt{1-\lambda}\,(x-s)\} \, ds & (\lambda < 1) \\[2ex] \displaystyle\int_0^x f(s)(x-s) \, ds & (\lambda = 1) \\[2ex] \dfrac{\lambda}{\sqrt{\lambda-1}} \displaystyle\int_0^x f(s) \sinh\{\sqrt{\lambda-1}\,(x-s)\} \, ds & (\lambda > 1) \end{cases}$$

$$(75)$$

3.4   Fredholm equation of the first kind

Here we are concerned with Fredholm equations of the form

$$f(x) = \int_{-\infty}^{\infty} k(x-s)\phi(s)\, ds \tag{76}$$

with an integral of the convolution type on the right-hand side. To solve this equation we take Fourier transforms. On applying the convolution theorem (13) this yields

$$F(u) = \sqrt{2\pi}\, K(u)\Phi(u) \tag{77}$$

which leads to the formal solution of (76):

$$\phi(x) = \frac{1}{\sqrt{2\pi}} \int_{-\infty}^{\infty} e^{-ixu}\Phi(u)\, du$$

$$= \frac{1}{2\pi} \int_{-\infty}^{\infty} e^{-ixu}\frac{F(u)}{K(u)}\, du. \tag{78}$$

3.4.1   Stieltjes integral equation

An interesting example of a Fredholm equation of the first kind possessing a convolution type integral is obtained by applying Laplace's integral operator

$$\int_{0}^{\infty} e^{-xs}\, ds$$

twice. Thus consider the equation

$$g(x) = \lambda \int_{0}^{\infty} e^{-xt}\, dt \int_{0}^{\infty} e^{-ts}\psi(s)\, ds. \tag{79}$$

Then, on reversing the order of the integrations, we see that

$$g(x) = \lambda \int_{0}^{\infty} \psi(s)\, ds \int_{0}^{\infty} e^{-(x+s)t}\, dt$$

$$= \lambda \int_{0}^{\infty} \frac{\psi(s)}{x+s}\, ds \tag{80}$$

which is known as *Stieltjes integral equation*.

Now setting

$$x = e^{\xi},\ s = e^{\sigma},\ e^{1/2\xi}g(e^{\xi}) = f(\xi),\ e^{1/2\xi}\psi(e^{\xi}) = \phi(\xi) \tag{81}$$

we obtain

$$f(\xi) = \frac{\lambda}{2} \int_{-\infty}^{\infty} \frac{\phi(\sigma)}{\cosh \frac{1}{2}(\xi - \sigma)} \, d\sigma. \tag{82}$$

This is a Fredholm equation of the first kind with kernel

$$k(\xi) = \frac{\lambda}{2 \cosh \frac{1}{2}\xi}. \tag{83}$$

To solve (82) we require to find the Fourier transform of $k(\xi)$ which is given by

$$\begin{aligned} K(u) &= \frac{\lambda}{\sqrt{2\pi}} \int_{-\infty}^{\infty} \frac{e^{iu\xi}}{e^{1/2\xi} + e^{-1/2\xi}} \, d\xi \\ &= \frac{\lambda}{\sqrt{2\pi}} \int_{0}^{\infty} \frac{x^{iu-(1/2)}}{1+x} \, dx \end{aligned} \tag{84}$$

on putting $x = e^{\xi}$. But it is well known that

$$\int_{0}^{\infty} \frac{x^{a-1}}{1+x} \, dx = \frac{\pi}{\sin \pi a} \tag{85}$$

and so we obtain

$$K(u) = \frac{\lambda \sqrt{\frac{\pi}{2}}}{\sin \{\pi(iu + \frac{1}{2})\}} = \frac{\lambda \sqrt{\frac{\pi}{2}}}{\cos i\pi u} = \frac{\lambda \sqrt{\frac{\pi}{2}}}{\cosh \pi u} \tag{86}$$

Now, making use of the solution (78) to the equation (76), we see that the solution of our integral equation (82) may be written as

$$\begin{aligned} \phi(\xi) &= \frac{1}{\lambda \pi \sqrt{2\pi}} \int_{-\infty}^{\infty} F(u) \cosh \pi u \, e^{-i\xi u} \, du \\ &= \frac{1}{2\pi\lambda\sqrt{2\pi}} \int_{-\infty}^{\infty} F(u)\{e^{-i(\xi + i\pi)u} + e^{-i(\xi - i\pi)u}\} \, du \end{aligned} \tag{87}$$

where $F(u)$ is the Fourier transform of $f(\xi)$. Thus

$$\phi(\xi) = \frac{1}{2\pi\lambda} \{f(\xi + i\pi) + f(\xi - i\pi)\} \tag{88}$$

and hence the solution to Stieltjes integral equation (80) is

$$\psi(x) = \frac{i}{2\pi\lambda} \{g(xe^{i\pi}) - g(xe^{-i\pi})\}. \tag{89}$$

Example. Suppose that

$$g(x) = \begin{cases} \dfrac{\ln \dfrac{x}{a}}{x-a} & (x \neq a) \\[3mm] \dfrac{1}{a} & (x = a) \end{cases} \tag{90}$$

Then (89) provides the solution

$$\psi(x) = \frac{1}{\lambda(x+a)} \tag{91}$$

to Stieltjes equation (80).

3.5   Volterra equation of the first kind

We now discuss Volterra equations of the form

$$f(x) = \int_0^x k(x-s)\phi(s)\,\mathrm{d}s \qquad (x > 0), \tag{92}$$

where $f(0) = 0$ and the integral on the right-hand side is of the convolution type again. This can be obtained from the Fredholm equation (76) of the first kind, examined in the previous section, by taking $f(x) = 0$, $k(x) = 0$ and $\phi(x) = 0$ for $x < 0$.

To solve equation (92) we introduce the Laplace transforms $F(u)$, $K(u)$ and $\Phi(u)$ of $f(x)$, $k(x)$ and $\phi(x)$. Then employing the convolution theorem for Laplace transforms we obtain

$$F(u) = K(u)\Phi(u) \tag{93}$$

and so

$$\phi(x) = \boldsymbol{L}^{-1}\left\{ \frac{F(u)}{K(u)} \right\} \tag{94}$$

where $\boldsymbol{L}^{-1}$ denotes the inverse of the Laplace transform.

Example. Consider the integral equation

$$f(x) = \int_0^x J_0(x-s)\phi(s)\,\mathrm{d}s \tag{95}$$

with a kernel of the convolution type given by

$$k(x) = J_0(x) \tag{96}$$

where $J_0(x)$ is the zero order Bessel function satisfying $J_0(0) = 1$. We can solve this equation by taking Laplace transforms. Since

$$L\{J_0(x)\} = \frac{1}{\sqrt{u^2 + 1}} \tag{97}$$

it follows that

$$\Phi(u) = \sqrt{u^2 + 1} \, F(u) \tag{98}$$

using (93). For the special case

$$f(x) = \sin x \tag{99}$$

we have

$$F(u) = \frac{1}{u^2 + 1} \tag{100}$$

and then

$$\Phi(u) = \frac{1}{\sqrt{u^2 + 1}}. \tag{101}$$

We see at once from (97) that for this case

$$\phi(x) = J_0(x) \tag{102}$$

which produces the interesting formula

$$\int_0^x J_0(x - s) J_0(s) \, \mathrm{d}s = \sin x. \tag{103}$$

3.5.1 Abel's integral equation

The equation solved by Abel has the form

$$f(x) = \int_0^x (x - s)^{-\alpha} \phi(s) \, \mathrm{d}s \qquad (0 < \alpha < 1) \tag{104}$$

with $f(0) = 0$, which is a Volterra integral equation of the first kind with a singular kernel of the convolution type given by

$$k(x) = x^{-\alpha}. \tag{105}$$

This equation is of considerable historical importance since for $\alpha = \frac{1}{2}$ it describes the tautochrone problem introduced briefly in section 1.1.

Now we have

$$K(u) = \int_0^\infty e^{-ux} x^{-\alpha} \, dx$$
$$= \Gamma(1-\alpha) u^{\alpha-1} \tag{106}$$

where $\Gamma(p)$ is the gamma function defined by

$$\Gamma(p) = \int_0^\infty e^{-x} x^{p-1} \, dx \qquad (p > 0), \tag{107}$$

so that

$$\mathbf{L}\{\phi(x)\} = \frac{F(u) u^{1-\alpha}}{\Gamma(1-\alpha)} \tag{108}$$

using (93).

Let us set

$$\phi(x) = \psi'(x) \tag{109}$$

where $\psi(0) = 0$, so that

$$\mathbf{L}\{\phi(x)\} = u \mathbf{L}\{\psi(x)\} \tag{110}$$

on performing a single integration by parts. Then

$$\mathbf{L}\{\psi(x)\} = \frac{F(u) u^{-\alpha}}{\Gamma(1-\alpha)} \tag{111}$$

and since we may write

$$u^{-\alpha} = \frac{\mathbf{L}(x^{\alpha-1})}{\Gamma(\alpha)},$$

it follows from the convolution theorem for Laplace transforms that

$$\psi(x) = \{\Gamma(\alpha)\Gamma(1-\alpha)\}^{-1} \int_0^x (x-s)^{\alpha-1} f(s) \, ds. \tag{112}$$

Using the well known result

$$\Gamma(\alpha)\Gamma(1-\alpha) = \frac{\pi}{\sin \pi\alpha} \tag{113}$$

we see that the solution of Abel's integral equation (104) is

$$\phi(x) = \frac{\sin \pi\alpha}{\pi} \frac{d}{dx} \int_0^x (x-s)^{\alpha-1} f(s) \, ds. \tag{114}$$

For the particular case $\alpha = \frac{1}{2}$ corresponding to the tautochrone problem this solution simplifies to

$$\phi(x) = \frac{1}{\pi} \frac{\mathrm{d}}{\mathrm{d}x} \int_0^x \frac{f(s)}{(x-s)^{1/2}} \, \mathrm{d}s. \tag{115}$$

3.6 Fox's integral equation

We seek the solution of the Fredholm integral equation of the second kind having the form

$$\phi(x) = f(x) + \int_0^\infty k(xs)\phi(s) \, \mathrm{d}s \qquad (0 < x < \infty) \tag{116}$$

named after Fox. This can be achieved by employing the Mellin transforms $\Phi(u)$, $F(u)$ and $K(u)$ of $\phi(x), f(x)$ and $k(x)$ respectively defined by (7).

Now we have shown in section 3.1 that $K(u)\Phi(1-u)$ is the Mellin transform of

$$\int_0^\infty k(xs)\phi(s) \, \mathrm{d}s$$

and so we see that

$$\Phi(u) = F(u) + K(u)\Phi(1-u) \tag{117}$$

on taking the Mellin transforms of both sides of (116). Also, replacing u by $1-u$, we have

$$\Phi(1-u) = F(1-u) + K(1-u)\Phi(u) \tag{118}$$

which enables us to write

$$\Phi(u) = \frac{F(u) + K(u)F(1-u)}{1 - K(u)K(1-u)} \tag{119}$$

Thus we have obtained a solution of Fox's integral equation (116) provided we can derive $\phi(x)$ from its Mellin transform $\Phi(u)$.

Example. Let us take as an example

$$k(x) = \lambda \sqrt{\frac{2}{\pi}} \sin x \tag{120}$$

Then the Mellin transform of $k(x)$ is given by

$$K(u) = \lambda \sqrt{\frac{2}{\pi}} \int_0^\infty x^{u-1} \sin x \, dx$$

$$= \lambda \sqrt{\frac{2}{\pi}} \Gamma(u) \sin \frac{\pi u}{2} \qquad (121)$$

and

$$K(u)K(1-u) = \frac{2\lambda^2}{\pi} \Gamma(u)\Gamma(1-u) \sin \frac{\pi u}{2} \cos \frac{\pi u}{2}$$

$$= \lambda^2 \Gamma(u)\Gamma(1-u) \frac{\sin \pi u}{\pi}$$

$$= \lambda^2 \qquad (122)$$

using (113). Hence, provided $\lambda^2 \neq 1$, we see that

$$\Phi(u) = \frac{F(u)}{1-\lambda^2} + \frac{\lambda}{1-\lambda^2} \sqrt{\frac{2}{\pi}} \Gamma(u) \sin \frac{\pi u}{2} F(1-u) \qquad (123)$$

making use of (119). But by the convolution theorem for Mellin transforms established in section 3.1 we know that $\Gamma(u) \sin \frac{\pi u}{2} F(1-u)$ is the Mellin transform of $\int_0^\infty \sin xs f(s) \, ds$, and so the solution of Fox's integral equation with kernel given by (120) is

$$\phi(x) = \frac{f(x)}{1-\lambda^2} + \frac{\lambda}{1-\lambda^2} \sqrt{\frac{2}{\pi}} \int_0^\infty \sin xs f(s) \, ds \qquad (0 < x < \infty)$$

$$(124)$$

This result can be verified directly by using Fourier's reciprocal sine formulae (1.2) and (1.3) in the form

$$f(x) = \frac{2}{\pi} \int_0^\infty \sin xs \, ds \int_0^\infty \sin st f(t) \, dt \qquad (0 < x < \infty) \qquad (125)$$

Problems

1. Solve the Lalesco-Picard equation

$$\phi(x) = \cos \mu x + \lambda \int_{-\infty}^\infty e^{-|x-s|} \phi(s) \, ds \qquad (\lambda < \tfrac{1}{2})$$

2. Find the solutions to the Volterra equations of the second kind

$$\phi(x) = f(x) \pm \int_0^x (x-s)\phi(s)\,ds$$

when (i) $f(x) = 1$, (ii) $f(x) = x$ using the general solution (65), and verify that they agree with the solutions obtained by solving the equivalent differential equations (see problem 3 at the end of chapter 2).

3. Find the solutions to the Volterra equation of the second kind

$$\phi(x) = f(x) + \lambda \int_0^x \sin(x-s)\phi(s)\,ds$$

when (i) $f(x) = x$ and $\lambda = 1$,
    (ii) $f(x) = e^{-x}$ and $\lambda = 2$,

using the general solution (75) and verify that they agree with the solutions obtained by solving the equivalent differential equations (see problem 4 at the end of chapter 2).

4. If $F(u)$ is the Fourier transform of $f(x)$ show that the transform of $f''(x)$ is $-u^2 F(u)$ provided $f, f' \to 0$ as $x \to \pm\infty$. Hence show that

$$f(x) = \int_{-\infty}^{\infty} e^{-|x-s|}\phi(s)\,ds \qquad (-\infty < x < \infty)$$

has the solution

$$\phi(x) = \tfrac{1}{2}\{f(x) - f''(x)\}$$

5. If $F(u)$ is the Laplace transform of $f(x)$ show that the transform of $f'(x)$ is $uF(u) - f(0)$ for $u > \alpha$ provided $f(x)e^{-\alpha x} \to 0$ as $x \to \infty$. Hence find the solution of the Volterra equation of the first kind

$$f(x) = \int_0^x e^{x-s}\phi(s)\,ds$$

where $f(0) = 0$.

6. Use Fourier's sine formulae to verify that the solution of Fox's integral equation

$$\phi(x) = f(x) + \lambda\sqrt{\frac{2}{\pi}}\int_0^{\infty} \sin xs\,\phi(s)\,ds \qquad (0 < x < \infty)$$

is given by (124).

7. Use the method of Mellin transforms to solve Fox's integral equation when the kernel is

$$k(x) = \lambda \sqrt{\frac{2}{\pi}} \cos x.$$

Verify the correctness of your solution by using Fourier's cosine formulae.

4

Method of successive approximations

4.1 Neumann series

A valuable method for solving integral equations of the second kind is based on an iterative procedure which yields a sequence of approximations leading to an infinite series solution associated with the names of Liouville and Neumann. It is sometimes called the Liouville-Neumann series but more often it is called the Neumann series.

Let us first examine the Fredholm equation

$$\phi(x) = f(x) + \lambda \int_a^b K(x, s)\phi(s) \, ds \qquad (a \le x \le b) \tag{1}$$

and consider the set of successive approximations to the solution $\phi(x)$ given by

$$\phi_0(x) = \phi^{(0)}(x)$$
$$\phi_1(x) = \phi^{(0)}(x) + \lambda\phi^{(1)}(x)$$
$$\phi_2(x) = \phi^{(0)}(x) + \lambda\phi^{(1)}(x) + \lambda^2\phi^{(2)}(x) \tag{2}$$

and so on, the Nth approximation being the sum

$$\phi_N(x) = \sum_{n=0}^{N} \lambda^n \phi^{(n)}(x) \qquad (N = 0, 1, 2, \ldots) \tag{3}$$

where

$$\phi^{(0)}(x) = f(x),$$
$$\phi^{(1)}(x) = \int_a^b K(x, s)\phi^{(0)}(s) \, ds,$$
$$\phi^{(2)}(x) = \int_a^b K(x, s)\phi^{(1)}(s) \, ds, \tag{4}$$

and in general

$$\phi^{(n)}(x) = \int_a^b K(x, s)\phi^{(n-1)}(s) \, ds \qquad (n \ge 1). \tag{5}$$

We see that the sequence of approximations is generated by an iterative process of successive substitutions on the right-hand side of (1).

Suppose that $f(x)$ and $K(x, s)$ are continuous functions in the range of definition so that they are bounded and we may write

$$|f(x)| \leq m \qquad (a \leq x \leq b)$$
$$|K(x, s)| \leq M \qquad (a \leq x \leq b, a \leq s \leq b) \qquad (6)$$

when m and M are positive constants. Then

$$|\phi^{(0)}(x)| = |f(x)| \leq m,$$

$$|\phi^{(1)}(x)| \leq \int_a^b |\phi^{(0)}(s)| \, |K(x, s)| \, ds \leq mM(b - a),$$

$$|\phi^{(2)}(x)| \leq \int_a^b |\phi^{(1)}(s)| \, |K(x, s)| \, ds \leq mM^2(b - a)^2,$$

and in general

$$|\phi^{(n)}(x)| \leq mM^n(b - a)^n. \qquad (7)$$

Hence

$$\left| \sum_{n=0}^{N} \lambda^n \phi^{(n)}(x) \right| \leq \sum_{n=0}^{N} |\lambda|^n |\phi^{(n)}(x)| \leq m \sum_{n=0}^{\infty} |\lambda|^n M^n (b - a)^n \qquad (8)$$

and so $\sum_{n=0}^{\infty} \lambda^n \phi^{(n)}(x)$ is absolutely and uniformly convergent in $a \leq x \leq b$ if $\rho = |\lambda| M(b - a) < 1$, that is if

$$|\lambda| < \frac{1}{M(b - a)} \qquad (9)$$

since then the series is dominated by the convergent geometric series

$$m \sum_{n=0}^{\infty} \rho^n = \frac{m}{1 - \rho}. \qquad (10)$$

When condition (9) is satisfied,

$$\phi(x) = \sum_{n=0}^{\infty} \lambda^n \phi^{(n)}(x) \qquad (11)$$

is a continuous solution of the integral equation (1).

The error made in replacing the exact solution ϕ by the Nth approximation ϕ_N is given by

$$r_N(x) = \phi(x) - \phi_N(x) = \sum_{n=N+1}^{\infty} \lambda^n \phi^{(n)}(x). \tag{12}$$

It is readily seen that

$$|r_N(x)| \leq m \sum_{n=N+1}^{\infty} \rho^n = \frac{m\rho^{N+1}}{1-\rho} \tag{13}$$

and so $|r_N(x)| \to 0$ as $N \to \infty$ if $\rho < 1$.

Now let us turn our attention to the Volterra equation

$$\phi(x) = f(x) + \lambda \int_a^x K(x, s)\phi(s) \, ds. \tag{14}$$

Then the previous analysis holds with $K(x, s) = 0$ for $s > x$. But we have further that

$$|\phi^{(0)}(x)| = |f(x)| \leq m,$$

$$|\phi^{(1)}(x)| \leq \int_a^x |\phi^{(0)}(s)| \, |K(x, s)| \, ds \leq mM \int_a^x ds = mM(x-a),$$

$$|\phi^{(2)}(x)| \leq \int_a^x |\phi^{(1)}(s)| \, |K(x, s)| \, ds \leq mM^2 \int_a^x (s-a) \, ds = \frac{mM^2(x-a)^2}{2!},$$

$$|\phi^{(3)}(x)| \leq \int_a^x |\phi^{(2)}(s)| \, |K(x, s)| \, ds \leq \frac{mM^3}{2!} \int_a^x (s-a)^2 \, ds = \frac{mM^3(x-a)^3}{3!}$$

and in general, assuming that

$$|\phi^{(n-1)}(x)| \leq \frac{mM^{n-1}(x-a)^{n-1}}{(n-1)!}$$

and using the principle of induction, we have

$$|\phi^{(n)}(x)| \leq \frac{mM^n}{(n-1)!} \int_a^x (s-a)^{n-1} \, ds = \frac{mM^n(x-a)^n}{n!}. \tag{15}$$

Hence

$$\left| \sum_{n=0}^{N} \lambda^n \phi^{(n)}(x) \right| \leq m \sum_{n=0}^{N} \frac{|\lambda|^n M^n (b-a)^n}{n!} \tag{16}$$

and so $\sum_{n=0}^{\infty} \lambda^n \phi^{(n)}(x)$ is absolutely and uniformly convergent in $a \leq x \leq b$ for all values of λ since it is dominated by

$$m \sum_{n=0}^{\infty} \frac{|\lambda|^n M^n (b-a)^n}{n!} = m \exp\{|\lambda| M(b-a)\}. \tag{17}$$

Thus the Neumann series (11) converges for all values of λ in the case of the Volterra equation (14) whereas it converges only for sufficiently small values of λ in the case of the Fredholm equation (1).

4.2   Iterates and the resolvent kernel

In the previous section we obtained solutions to the Fredholm and Volterra equations of the second kind in the form of the infinite series (11). It is convenient to express these solutions in terms of *iterated kernels* defined by

$$K_1(x, s) = K(x, s) \tag{18}$$

$$K_n(x, s) = \int_a^b K(x, t) K_{n-1}(t, s)\, dt \qquad (n \geq 2) \tag{19}$$

so that

$$K_2(x, s) = \int_a^b K(x, t_1) K(t_1, s)\, dt_1,$$

$$K_3(x, s) = \int_a^b K(x, t_1) K_2(t_1, s)\, dt_1 \tag{20}$$

$$= \int_a^b \int_a^b K(x, t_1) K(t_1, t_2) K(t_2, s)\, dt_1\, dt_2 \tag{21}$$

and in general

$$K_n(x, s) = \int_a^b \cdots \int_a^b K(x, t_1) K(t_1, t_2) \cdots$$

$$K(t_{n-2}, t_{n-1}) K(t_{n-1}, s)\, dt_1 \cdots dt_{n-1} \tag{22}$$

It follows at once that the iterated kernels satisfy

$$K_n(x, s) = \int_a^b K_p(x, t) K_q(t, s)\, dt \tag{23}$$

for any p and q with $p + q = n$.

Now

$$\phi^{(n)}(x) = \int_a^b K_n(x, s) f(s)\, ds \tag{24}$$

and hence we may write the solution (11) of the Fredholm equation (1) in the form

$$\phi(x) = f(x) + \sum_{n=1}^{\infty} \lambda^n \int_a^b K_n(x, s) f(s) \, ds \qquad (25)$$

or equivalently

$$\phi(x) = f(x) + \lambda \int_a^b R(x, s; \lambda) f(s) \, ds \qquad (26)$$

where

$$R(x, s; \lambda) = \sum_{n=0}^{\infty} \lambda^n K_{n+1}(x, s) \qquad (27)$$

is the solving kernel or *resolvent kernel* already introduced in section 1.3.3 during the discussion of separable kernels.

If $f(x)$ and $K(x, s)$ are continuous functions satisfying (6) and $\rho = |\lambda| M(b - a) < 1$, the infinite series (27) for $R(x, s; \lambda)$ is absolutely and uniformly convergent in $a \leqslant x \leqslant b$, $a \leqslant s \leqslant b$ since

$$\left| \sum_{n=0}^{N} \lambda^n K_{n+1}(x, s) \right| \leqslant \sum_{n=0}^{N} |\lambda|^n |K_{n+1}(x, s)|$$

$$\leqslant \sum_{n=0}^{\infty} |\lambda|^n M^{n+1}(b - a)^n = \frac{M}{1 - \rho}.$$

Moreover for Volterra kernels satisfying $K(x, s) = 0$ for $x < s$ we have, using (6) and the knowledge that the integrand on the right-hand side of (22) vanishes except for $x \geqslant t_1 \geqslant t_2 \geqslant \cdots \geqslant t_{n-1} \geqslant s$:

$$|K_n(x, s)| \leqslant \frac{M^n (x - s)^{n-1}}{(n - 1)!} \leqslant \frac{M^n (b - a)^{n-1}}{(n - 1)!}$$

and so the infinite series (27) for $R(x, s; \lambda)$ is absolutely and uniformly convergent in $a \leqslant x \leqslant b$, $a \leqslant s \leqslant b$ for all values of λ since

$$\left| \sum_{n=0}^{N} \lambda^n K_{n+1}(x, s) \right| \leqslant \sum_{n=0}^{N} |\lambda|^n |K_{n+1}(x, s)|$$

$$\leqslant M \sum_{n=0}^{\infty} \frac{|\lambda|^n M^n (b - a)^n}{n!}$$

$$= M \exp \{ |\lambda| M(b - a) \}.$$

It can also be shown that the resolvent kernel is continuous in the same region for $|\lambda| < \{ M(b - a) \}^{-1}$ in the case of continuous

Fredholm kernels, and for all λ in the case of continuous Volterra kernels.

Example 1.   As a first example of the Neumann series we consider the Fredholm equation of the second kind

$$\phi(x) = f(x) + \lambda \int_a^b u(x)\overline{v(s)}\phi(s)\,ds \qquad (28)$$

with separable kernel $K(x, s) = u(x)\overline{v(s)}$. This equation was discussed previously in section 1.3.3 where an exact solution (1.30) in closed analytical form was obtained.

Now we have

$$K_1(x, s) = K(x, s) = u(x)\overline{v(s)}$$

$$K_2(x, s) = \int_a^b K_1(x, t)K_1(t, s)\,dt$$

$$= u(x)\overline{v(s)}\int_a^b \overline{v(t)}u(t)\,dt$$

$$= \alpha u(x)\overline{v(s)},$$

where

$$\alpha = \int_a^b \overline{v(t)}u(t)\,dt,$$

and if we assume that

$$K_n(x, s) = \alpha^{n-1}u(x)\overline{v(s)}$$

then

$$K_{n+1}(x, s) = \int_a^b K_1(x, t)K_n(t, s)\,dt$$

$$= \alpha^{n-1}u(x)\overline{v(s)}\int_a^b \overline{v(t)}u(t)\,dt$$

$$= \alpha^n u(x)\overline{v(s)} \qquad (29)$$

establishing the general form of the interated kernel $K_{n+1}(x, s)$ by the principle of induction.

Hence the resolvent kernel is

$$R(x, s; \lambda) = \sum_{n=0}^{\infty} \lambda^n K_{n+1}(x, s)$$

$$= u(x)\overline{v(s)} \sum_{n=0}^{\infty} (\lambda\alpha)^n \tag{30}$$

$$= \frac{u(x)\overline{v(s)}}{1 - \lambda\alpha} \tag{31}$$

provided $|\lambda\alpha| < 1$. This is in accordance with the exact solution (1.32) derived previously for all $\lambda \neq \alpha^{-1}$.

The homogeneous equation corresponding to (28) is

$$\phi(x) = \lambda \int_a^b u(x)\overline{v(s)}\phi(s)\,ds$$

It possesses one characteristic value $\lambda_1 = \alpha^{-1}$ and we see that the Neumann expansion (30) converges to (31) for $|\lambda| < |\lambda_1|$.

Example 2.   We now consider the Fredholm equation

$$\phi(x) = 1 + \lambda \int_0^1 xs\phi(s)\,ds \qquad (0 \leqslant x \leqslant 1) \tag{32}$$

with $f(x) = 1$ and separable kernel $K(x, s) = xs$. This was examined earlier as example 1 of section 1.3.3.

We have

$$\alpha = \int_0^1 t^2\,dt = \tfrac{1}{3} \tag{33}$$

so that

$$K_{n+1}(x, s) = \frac{xs}{3^n}. \tag{34}$$

Hence the resolvent kernel is

$$R(x, s; \lambda) = \sum_{n=0}^{\infty} \lambda^n K_{n+1}(x, s)$$

$$= xs \sum_{n=0}^{\infty} \left(\frac{\lambda}{3}\right)^n$$

$$= \frac{xs}{1 - \lambda/3} \tag{35}$$

provided $|\lambda| < 3$, this being less severe that the condition given by $\rho < 1$.

Thus, using (26) we see that

$$\phi(x) = f(x) + \lambda \int_0^1 R(x, s; \lambda) f(s) \, ds$$

$$= 1 + \frac{\lambda x}{1 - \lambda/3} \int_0^1 s \, ds$$

$$= 1 + \frac{3\lambda x}{2(3 - \lambda)} \tag{36}$$

which is in accordance with the exact solution (1.38) derived previously for all $\lambda \neq 3$.

Example 3. Lastly we discuss the Volterra equation with kernel

$$K(x, s) = \begin{cases} e^{x-s} & (x > s) \\ 0 & (s > x) \end{cases} \tag{37}$$

This kernel is of the convolution type and was discussed before as example 2 of section 3.3 where an exact solution in closed analytical form was derived.

Now

$$K_2(x, s) = \int_s^x K(x, t) K(t, s) \, dt$$

$$= \int_s^x e^{x-t} \cdot e^{t-s} \, dt$$

$$= e^{x-s} \int_s^x dt$$

$$= e^{x-s}(x - s) \qquad (x > s),$$

$$K_3(x, s) = \int_s^x K(x, t) K_2(t, s) \, dt$$

$$= \int_s^x e^{x-t} \cdot e^{t-s}(t - s) \, dt$$

$$= e^{x-s} \int_s^x (t - s) \, dt$$

$$= e^{x-s} \frac{(x - s)^2}{2!} \qquad (x > s)$$

and in general, assuming that

$$K_n(x, s) = e^{x-s} \frac{(x-s)^{n-1}}{(n-1)!} \qquad (x > s)$$

we have

$$
\begin{aligned}
K_{n+1}(x, s) &= \int_s^x K(x, t) K_n(t, s) \, dt \\
&= \int_s^x e^{x-t} \cdot e^{t-s} \frac{(t-s)^{n-1}}{(n-1)!} \, dt \\
&= e^{x-s} \int_s^x \frac{(t-s)^{n-1}}{(n-1)!} \, dt \\
&= e^{x-s} \frac{(x-s)^n}{n!} \qquad (x > s)
\end{aligned}
\tag{38}
$$

which establishes the general form of the iterated kernel by the principle of induction.

Hence the resolvent kernel given by (27) is

$$
\begin{aligned}
R(x, s; \lambda) &= \sum_{n=0}^{\infty} \frac{\lambda^n (x-s)^n}{n!} e^{x-s} \\
&= e^{(\lambda+1)(x-s)} \qquad (x > s)
\end{aligned}
\tag{39}
$$

for all λ, which is in agreement with the solution (70) obtained in section 3.3 since the resolvent kernel vanishes for $x < s$.

Problems

1. Use the method of successive approximations to solve

$$\phi(x) = 1 + \lambda \int_0^1 e^{x+s} \phi(s) \, ds$$

and verify that your solution agrees with the solution to problem 3 at the end of chapter 1 for $\lambda < 2/(e^2 - 1)$.

2. Use the method of successive approximations to solve

$$\phi(x) = x + \lambda \int_0^{\pi} \sin nx \, \sin ns \phi(s) \, ds$$

where n is an integer, and verify that your solution agrees with the solution to problem 5 at the end of chapter 1 for $\lambda < 2/\pi$.

3. Obtain the solution in closed analytical form of

$$\phi(x) = f(x) + \int_0^{2\pi} K(x, s)\phi(s)\, ds$$

where

$$K(x, s) = \sum_{n=1}^{\infty} a_n \sin nx \cos ns$$

and

$$\sum_{n=1}^{\infty} |a_n| < \infty,$$

using the method of successive approximations.

4. Use the method of successive approximations to solve

$$\phi(x) = 1 + \lambda \int_0^x \phi(s)\, ds,$$

verifying that your solution for $\lambda = 1$ agrees with that obtained to problem 1(i) at the end of chapter 2.

5. Use the method of successive approximations to obtain the resolvent kernel for

$$\phi(x) = f(x) + \lambda \int_0^x (x - s)\phi(s)\, ds$$

Verify that this agrees with the solutions to problem 3 at the end of chapter 2.

6. Use the method of successive approximations to show that the resolvent kernel for

$$\phi(x) = 1 + \lambda \int_0^x xs\phi(s)\, ds$$

is given by

$$R(x, s; \lambda) = xs \sum_{n=0}^{\infty} \left(\frac{\lambda}{3}\right)^n \frac{(x^3 - s^3)^n}{n!}$$

and hence solve the integral equation.

Integral equations
with singular kernels

5.1 Generalization to higher dimensions

So far in this book we have been concerned solely with integral
equations which involve an unknown function $\phi(x)$ of a single real
variable x. However it is interesting to consider integral equations in
higher dimensions and to this end we suppose that V is a region of
an n-dimensional space and let M, N denote points in the region V.
Then an integral equation of the second kind takes the form

$$\phi(M) = f(M) + \int_V K(M, N)\phi(N) \, dv_N \tag{1}$$

where $K(M, N)$ is the kernel, $f(M)$ is a given function, $\phi(M)$ is the
function which we require to find, and the integration in (1) is over
all the points N of the n-dimensional region V.

Also an integral equation of the first kind takes the form

$$f(M) = \int_V K(M, N)\phi(N) \, dv_N. \tag{2}$$

A common type of singular kernel is given by

$$K(M, N) = \frac{F(M, N)}{r^\alpha} \tag{3}$$

where $F(M, N)$ is a bounded function and r is the distance between
the points M and N in the n-dimensional space. This type of kernel
is called *polar* and as $r \to 0$ it approaches an infinite value. It is said
to give rise to an integral equation with a *weak* singularity if
$0 < \alpha < n$. The polar kernel (3) is square integrable, i.e.

$$\int_V \int_V |K(M, N)|^2 \, dv_M \, dv_N < \infty, \tag{4}$$

when α is restricted further to the range $0 < \alpha < n/2$, as we have
shown for the one-dimensional case ($n = 1$) in section 1.4.

5.2 Green's functions in two and three dimensions

We now consider the linear second order partial differential equation

$$L\psi = \Omega \tag{5}$$

and begin by investigating the two-dimensional case for which

$$L = \frac{\partial^2}{\partial x^2} + \frac{\partial^2}{\partial y^2} + q(x, y) \tag{6}$$

Then the corresponding Green's function $G(x, y; s, t)$ satisfies

$$LG(x, y; s, t) = \delta(x - s)\delta(y - t). \tag{7}$$

Integrating over the interior of a circle γ of radius ε centered at the point with coordinates s, t we have

$$\int_\gamma LG(x, y; s, t)\, dx\, dy = 1 \tag{8}$$

since the integral of the δ functions amounts to unity. This leads to

$$\lim_{\varepsilon \to 0} \int_\gamma \frac{\partial G}{\partial \rho}\, d\sigma = 1 \tag{9}$$

where $d\sigma$ represents an element of arc length and

$$\rho^2 = (x - s)^2 + (y - t)^2. \tag{10}$$

We now see that

$$2\pi\rho\frac{\partial G}{\partial \rho} \to 1 \tag{11}$$

as $\rho \to 0$ and so

$$G(x, y; s, t) = \frac{1}{2\pi}\ln \rho + g(x, y; s, t) \tag{12}$$

where

$$Lg(x, y; s, t) = -\frac{1}{2\pi}q(x, y)\ln \rho \tag{13}$$

and g is chosen so that G satisfies prescribed boundary conditions.

In the three-dimensional case

$$L = \nabla^2 + q(\mathbf{r}) \tag{14}$$

and the Green's function satisfies

$$LG(\mathbf{r}, \mathbf{s}) = \delta(\mathbf{r} - \mathbf{s}) \tag{15}$$

where $\mathbf{r}$ and $\mathbf{s}$ are the position vectors of the points with coordinates (x, y, z) and (s, t, u) respectively.

Integrating over the region contained by a sphere S of radius ε centered at the point with position vector $\mathbf{s}$ we obtain

$$\int_S LG(\mathbf{r}, \mathbf{s}) \, d\mathbf{r} = 1 \tag{16}$$

giving

$$\lim_{\varepsilon \to 0} \oint_S \frac{\partial G}{\partial R} \, dS = 1 \tag{17}$$

where the integration in (17) is over the surface of the sphere S and

$$R^2 = |\mathbf{r} - \mathbf{s}|^2 = (x - s)^2 + (y - t)^2 + (z - u)^2. \tag{18}$$

It follows that

$$4\pi R^2 \frac{\partial G}{\partial R} \to 1 \tag{19}$$

as $R \to 0$ and hence the Green's function has the form

$$G(\mathbf{r}, \mathbf{s}) = -\frac{1}{4\pi R} + g(\mathbf{r}, \mathbf{s}) \tag{20}$$

where

$$Lg(\mathbf{r}, \mathbf{s}) = \frac{q(\mathbf{r})}{4\pi R} \tag{21}$$

and we choose g so that G satisfies given boundary conditions.

5.3   Dirichlet's problem

The problem named after Dirichlet is concerned with determining a function ψ which has prescribed values f over the boundary Γ of a certain simply connected region R and which is harmonic, i.e., satisfies Laplace's equation, at all interior points of R.

We begin by considering the two-dimensional Dirichlet problem in which we have a plane region R bounded by a closed contour γ with a continuously turning tangent. Let us suppose that $\psi(x, y)$ is the harmonic function satisfying

$$\nabla_1^2 \psi \equiv \frac{\partial^2 \psi}{\partial x^2} + \frac{\partial^2 \psi}{\partial y^2} = 0 \qquad (22)$$

in R and having the prescribed values given by $f(s, t)$ at all points (s, t) of γ. We now express ψ as the real part of an analytic function $F(z)$ of the complex variable $z = x + iy$ by putting

$$\psi(x, y) = \operatorname{Re} F(z) \qquad (23)$$

and look for a solution in the form of the Cauchy-type integral

$$F(z) = \frac{1}{2\pi i} \int_\gamma \frac{\mu(\zeta)}{\zeta - z} \, d\zeta \qquad (24)$$

where the density $\mu(\zeta)$ is a real function of the complex variable ζ and the contour γ is directed in the anticlockwise sense. Next we allow z to approach a point w on the boundary curve γ from the interior of R. In the limit we obtain

$$F(w) = \tfrac{1}{2}\mu(w) + \frac{1}{2\pi i} P \int_\gamma \frac{\mu(\zeta)}{\zeta - w} \, d\zeta \qquad (25)$$

where the integral occurring on the right-hand side of (25) is its Cauchy principal value (indicated by the prefix P):

$$\lim_{\varepsilon \to 0} \int_{\gamma_\varepsilon} \frac{\mu(\zeta)}{\zeta - w} \, d\zeta \qquad (26)$$

where γ_ε is the part of γ outside a circle of radius ε centered at the point w, and the term $\tfrac{1}{2}\mu(w)$ is the contribution arising from the singularity at $\zeta = w$.

Now taking the real parts of (25) we find that

$$f(x, y) = \tfrac{1}{2}\mu(x, y) + \frac{1}{2\pi} P \int_\gamma \mu(\zeta) \operatorname{Im}\left(\frac{d\zeta}{\zeta - w}\right) \qquad (27)$$

where (x, y) are the coordinates of the point w on γ.

Let us write $\zeta - w = re^{i\theta}$. Then the imaginary part of $d\zeta/(\zeta - w)$

becomes

$$\operatorname{Im}\left(\frac{d\zeta}{\zeta - w}\right) = \operatorname{Im}[d\{\ln(\zeta - w)\}]$$

$$= \operatorname{Im}[d(\ln r + i\theta)]$$

$$= d\theta$$

$$= \frac{\partial \theta}{\partial \sigma} d\sigma \tag{28}$$

where $d\sigma$ is an element of arc of γ. Using one of the Cauchy-Riemann equations $\dfrac{\partial \psi}{\partial x} = \dfrac{\partial \phi}{\partial y}, \dfrac{\partial \psi}{\partial y} = -\dfrac{\partial \phi}{\partial x}$ connecting the real and imaginary parts ψ, ϕ of an analytic function we get

$$\frac{\partial \theta}{\partial \sigma} = \frac{\partial}{\partial n}(\ln r) = \frac{1}{r}\frac{\partial r}{\partial n} \tag{29}$$

where n is measured in the direction of the outward normal to γ (see Fig. 4).

Now

$$\frac{\partial r}{\partial n} = \hat{\mathbf{r}} \cdot \hat{\mathbf{n}} \tag{30}$$

where $\hat{}$ denotes a unit vector, so that we obtain

$$f(x, y) = \tfrac{1}{2}\mu(x, y) + \frac{1}{2\pi} P\int_\gamma \mu(s, t)\frac{\hat{\mathbf{r}} \cdot \hat{\mathbf{n}}}{r} d\sigma. \tag{31}$$

Since we can characterize the point ζ by its arc distance σ along γ from the point w, we may rewrite (31) in the one-dimensional form

$$\mu(\sigma) = 2f(\sigma) - \frac{1}{\pi} P\int_\gamma \mu(\sigma)\frac{\hat{\mathbf{r}} \cdot \hat{\mathbf{n}}}{r} d\sigma \tag{32}$$

which is a Fredholm integral equation of the second kind with a polar kernel (3) for $\alpha = 1$.

We next turn to the three-dimensional interior Dirichlet problem. Let S be the boundary surface of the three-dimensional region R and suppose that the solution ψ of Laplace's equation $\nabla^2 \psi = 0$ possesses the prescribed values at the boundary surface given by the function f.

If M is an interior point of R we may express the solution as the potential of a double layer, which in electricity is a distribution of

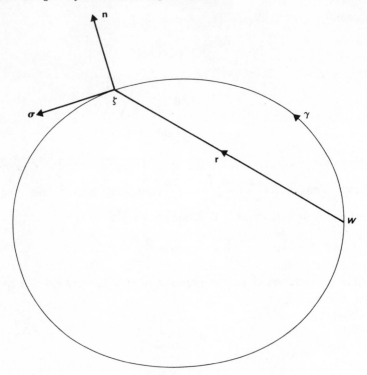

Fig. 4. Dirichlet's problem.

electric dipoles spread over the surface S with their axes in the directions of the normals. Thus we have

$$\psi(M) = \frac{1}{4\pi} \int_S \mu(N) \frac{\hat{\mathbf{r}} \cdot \hat{\mathbf{n}}}{r^2} \, dS_N \qquad (33)$$

where $\hat{\mathbf{n}}$ is a unit vector along the outward normal at the point N of the surface S, $\mu(N)$ is the density of the double layer and $\mathbf{r} = \overrightarrow{MN}$. Then allowing M to tend to a point of S from the interior of R yields the two-dimensional Fredholm integral equation of the second kind for μ:

$$\mu(M) = 2f(M) - \frac{1}{2\pi} P \int_S \mu(N) \frac{\hat{\mathbf{r}} \cdot \hat{\mathbf{n}}}{r^2} \, dS_N \qquad (34)$$

with a polar kernel (3) for $\alpha = 2$.

5.3.1 Poisson's formula for the unit disc

We now search for the solution $\psi(r, \theta)$ of Dirichlet's problem for a unit circle γ, the polar coordinates (r, θ) being referred to the centre of the circle. To this end we consider Laplace's equation in a plane

$$\nabla_1^2 \psi = 0 \tag{35}$$

where $\psi(r, \theta) \to f(\theta)$ as $r \to 1 - 0$.

If $G(\mathbf{r}, \mathbf{s})$ is the Green's function satisfying

$$\nabla_1^2 G(\mathbf{r}, \mathbf{s}) = \delta(\mathbf{r} - \mathbf{s}) \tag{36}$$

with $G(\mathbf{r}, \mathbf{s}) = 0$ when $\mathbf{s}$ is on the boundary given by $s = 1$, it follows from (12) that

$$G(\mathbf{r}, \mathbf{s}) = \frac{1}{2\pi} \{\ln |\mathbf{r} - \mathbf{s}| - \ln (r |\mathbf{r}' - \mathbf{s}|)\} \tag{37}$$

where $\mathbf{r}'$ is the *inverse point* of $\mathbf{r}$ with respect to the unit circle γ so that $rr' = 1$ and thus, using similar triangles (see Fig. 5), $|\mathbf{r} - \mathbf{s}| = r |\mathbf{r}' - \mathbf{s}|$ if $\mathbf{s}$ is on the boundary.

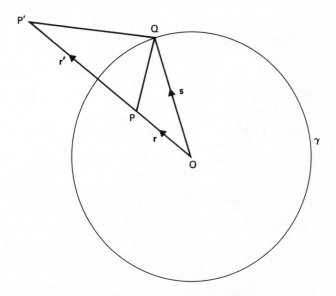

Fig. 5. Poisson's formula: P' is the inverse point of P with respect to the unit circle γ centered at O so that OP.OP' = OQ^2 = 1, and OQP and OP'Q are similar triangles.

Green's theorem states that

$$\int_R (\phi \nabla_1^2 \psi - \psi \nabla_1^2 \phi) \, dS = \int_\gamma \left(\phi \frac{\partial \psi}{\partial n} - \psi \frac{\partial \phi}{\partial n} \right) d\sigma \qquad (38)$$

where n is measured in the direction of the outward normal to the boundary curve γ and so, setting $\phi = G(\mathbf{r}, \mathbf{s})$, we obtain

$$\psi(\mathbf{r}) = \int_\gamma \psi(\mathbf{s}) \frac{\partial}{\partial n} G(\mathbf{r}, \mathbf{s}) \, d\sigma. \qquad (39)$$

Now

$$|\mathbf{r} - \mathbf{s}|^2 = r^2 + s^2 - 2rs \cos(\theta - \chi)$$

and

$$r^2 |\mathbf{r}' - \mathbf{s}|^2 = r^2 \{r'^2 + s^2 - 2r's \cos(\theta - \chi)\}$$

where (s, χ) are the polar coordinates of the point $\mathbf{s}$ on the unit circle γ. But $r' = r^{-1}$ and so

$$r^2 |\mathbf{r}' - \mathbf{s}|^2 = 1 + r^2 s^2 - 2rs \cos(\theta - \chi).$$

Hence

$$\psi(\mathbf{r}) = \frac{1}{2\pi} \int_\gamma \psi(\mathbf{s}) \frac{\partial}{\partial s} [\tfrac{1}{2} \ln \{r^2 + s^2 - 2rs \cos(\theta - \chi)\}$$
$$- \tfrac{1}{2} \ln \{1 + r^2 s^2 - 2rs \cos(\theta - \chi)\}] \, d\sigma$$

and since $s = 1$ and $\psi(\mathbf{s}) = f(\chi)$ over the unit circle γ we find that

$$\psi(\mathbf{r}) = \frac{1 - r^2}{2\pi} \int_0^{2\pi} f(\chi) [1 + r^2 - 2r \cos(\theta - \chi)]^{-1} \, d\chi \qquad (40)$$

which is Poisson's formula for the unit disc.

5.3.2 Poisson's formula for the half plane

We consider the half plane $y > 0$ bounded by the straight line $y = 0$ and suppose that $\psi(x, y) \to f(x)$ as $y \to +0$. Then the Green's function (12) which vanishes when the point $\mathbf{s}$ with coordinates (s, t) lies on the boundary line $t = 0$, takes the form

$$G(\mathbf{r}, \mathbf{s}) = \frac{1}{2\pi} \{\ln |\mathbf{r} - \mathbf{s}| - \ln |\mathbf{r}' - \mathbf{s}|\} \qquad (41)$$

where the coordinates of $\mathbf{r}$ and $\mathbf{r}'$ are (x, y) and $(x, -y)$ respectively.

Now using (39) we obtain

$$
\begin{aligned}
\psi(x, y) = -\frac{1}{2\pi} \int_{-\infty}^{\infty} f(s) \frac{\partial}{\partial t} [\tfrac{1}{2}\ln\{(s-x)^2 + (t-y)^2\} \\
-\tfrac{1}{2}\ln\{(s-x)^2 + (t+y)^2\}]_{t=0}\,ds \\
= \frac{y}{\pi} \int_{-\infty}^{\infty} f(s)[(s-x)^2 + y^2]^{-1}\,ds
\end{aligned}
\tag{42}
$$

which is Poisson's formula for the half plane.

5.3.3 Hilbert kernel

An integral equation with the singular kernel

$$
\cot\left(\frac{\chi - \theta}{2}\right)
\tag{43}
$$

named after Hilbert can be obtained by using Poisson's formula (40) for a unit disc and setting $z = re^{i\theta}$ and $\zeta = e^{i\chi}$. Then $d\zeta/\zeta = i\,d\chi$ and

$$
\frac{\zeta + z}{\zeta - z} = \frac{1 + r\cos(\theta - \chi) + ir\sin(\theta - \chi)}{1 - r\cos(\theta - \chi) - ir\sin(\theta - \chi)}
$$

giving

$$
\mathrm{Re}\left(\frac{\zeta + z}{\zeta - z}\right) = \frac{1 - r^2}{1 + r^2 - 2r\cos(\theta - \chi)}.
$$

Hence

$$
\psi(r, \theta) = \mathrm{Re}\,\frac{1}{2\pi i} \int_{\gamma} f(\chi) \frac{\zeta + z}{\zeta - z} \frac{d\zeta}{\zeta}
\tag{44}
$$

where γ is the circle $|\zeta| = 1$.

Let $\phi(r, \theta)$ be the harmonic function which is conjugate to $\psi(r, \theta)$. Then $\phi(r, \theta)$ is determined apart from an additive arbitrary constant which we choose so as to make $\phi(r, \theta)$ vanish at $r = 0$. We have then

$$
\psi(r, \theta) + i\phi(r, \theta) = \frac{1}{2\pi i} \int_{\gamma} f(\chi) \frac{\zeta + z}{\zeta - z} \frac{d\zeta}{\zeta}.
\tag{45}
$$

Now letting $r \to 1 - 0$ so that z approaches a point of γ from the interior of the unit disc, we obtain

$$
\psi(1, \theta) + i\phi(1, \theta) = f(\theta) + \frac{1}{2\pi i} P\!\int_{\gamma} f(\chi) \frac{\zeta + z}{\zeta - z} \frac{d\zeta}{\zeta}.
$$

We already know that $\psi(1, \theta) = f(\theta)$. Putting $\phi(1, \theta) = g(\theta)$ we find that

$$g(\theta) = \frac{1}{2\pi i} P \int_0^{2\pi} f(\chi) \frac{e^{i\chi} + e^{i\theta}}{e^{i\chi} - e^{i\theta}} \, d\chi$$

$$= -\frac{i}{2\pi} P \int_0^{2\pi} f(\chi) \frac{e^{i[(\chi - \theta)/2]} + e^{-i[(\chi - \theta)/2]}}{e^{i[(\chi - \theta)/2]} - e^{-i[(\chi - \theta)/2]}} \, d\chi$$

and so

$$g(\theta) = -\frac{1}{2\pi} P \int_0^{2\pi} f(\chi) \cot\left(\frac{\chi - \theta}{2}\right) d\chi \qquad (46)$$

which is an integral equation of the first kind for $f(\chi)$ having the Hilbert singular kernel (43).

Now we turn to the Poisson formula for the half plane and set $z = x + iy$ and $\zeta = s + it$. On the $t = 0$ axis we have

$$\text{Im}\left(\frac{1}{\zeta - z}\right) = \text{Im}\left(\frac{1}{s - x - iy}\right)$$

$$= \frac{y}{(s - x)^2 + y^2}$$

and so

$$\psi(x, y) = \text{Re} \, \frac{1}{\pi i} \int_\gamma \frac{\psi(\zeta)}{\zeta - z} \, d\zeta \qquad (47)$$

where γ is the straight line $t = 0$ together with the infinite semicircle in the upper half plane traced in the anticlockwise sense. We let $\phi(x, y)$ be the harmonic function conjugate to $\psi(x, y)$. Then

$$\psi(x, y) + i\phi(x, y) = \frac{1}{\pi i} \int_\gamma \frac{\psi(\zeta)}{\zeta - z} \, d\zeta \qquad (48)$$

and so, letting $y \to +0$, we obtain

$$\psi(x, +0) + i\phi(x, +0) = f(x) + \frac{1}{\pi i} P \int_{-\infty}^{\infty} \frac{f(s)}{s - x} \, ds$$

giving

$$g(x) = -\frac{1}{\pi} P \int_{-\infty}^{\infty} \frac{f(s)}{s - x} \, ds \qquad (49)$$

where $f(x) = \psi(x, +0)$, $g(x) = \phi(x, +0)$ and the integral on the right-hand side is the principal value defined by

$$\lim_{\varepsilon \to 0} \left[\int_{-\infty}^{x-\varepsilon} \frac{f(s)}{s-x} \, \mathrm{d}s + \int_{x+\varepsilon}^{\infty} \frac{f(s)}{s-x} \, \mathrm{d}s \right]. \tag{50}$$

Equation (49) is an integral equation of the first kind having the singular kernel $(s-x)^{-1}$.

5.3.4 Hilbert transforms

Suppose that $\psi(x, y)$, $\psi_1(x, y)$, $\psi_2(x, y)$ are functions which are harmonic in a plane region R bounded by a closed curve γ, and ψ_1 is conjugate to ψ while ψ_2 is conjugate to ψ_1. Then using the Cauchy-Riemann equations for the analytic functions $\psi + i\psi_1$ and $\psi_1 + i\psi_2$ respectively we obtain

$$\frac{\partial \psi}{\partial x} = \frac{\partial \psi_1}{\partial y}, \qquad \frac{\partial \psi}{\partial y} = -\frac{\partial \psi_1}{\partial x} \tag{51}$$

and

$$\frac{\partial \psi_1}{\partial x} = \frac{\partial \psi_2}{\partial y}, \qquad \frac{\partial \psi_1}{\partial y} = -\frac{\partial \psi_2}{\partial x} \tag{52}$$

so that

$$\frac{\partial}{\partial x}(\psi + \psi_2) = 0, \qquad \frac{\partial}{\partial y}(\psi + \psi_2) = 0 \tag{53}$$

Hence

$$\psi = -\psi_2 + C \tag{54}$$

where C is a constant.

Let us now apply this result to the case of the unit disc for which γ is the circle $r = 1$. We take the values of ψ, ψ_1, ψ_2 over γ to be $f(\theta)$, $f_1(\theta)$, $f_2(\theta)$ respectively and choose ψ_2 to vanish at $r = 0$ so that $C = \psi(r = 0)$. But Poisson's formula (40) for the unit disc informs us that

$$\psi(r = 0) = \frac{1}{2\pi} \int_0^{2\pi} f(\chi) \, \mathrm{d}\chi \tag{55}$$

and so

$$\psi = -\psi_2 + \frac{1}{2\pi} \int_0^{2\pi} f(\chi) \, \mathrm{d}\chi \tag{56}$$

giving in particular

$$f(\theta) = -f_2(\theta) + \frac{1}{2\pi} \int_0^{2\pi} f(\chi)\, d\chi \qquad (57)$$

Hence, if $f(\theta)$ and $g(\theta)$ are the real and imaginary parts of the analytic function $\Phi(z) = \psi + i\phi$ over the boundary circle $r = 1$, we see that

$$g(\theta) = -\frac{1}{2\pi} P \int_0^{2\pi} f(\chi) \cot\left(\frac{\chi - \theta}{2}\right) d\chi \qquad (58)$$

and

$$f(\theta) = \frac{1}{2\pi} P \int_0^{2\pi} g(\chi) \cot\left(\frac{\chi - \theta}{2}\right) d\chi + \frac{1}{2\pi} \int_0^{2\pi} f(\chi)\, d\chi \qquad (59)$$

These are the reciprocal formulae deduced by Hilbert in 1904. Equation (59) provides a solution of the integral equation (58) of the first kind with the singular kernel (43). Combining (58) and (59) together gives the *Hilbert formula*

$$f(\theta) - \frac{1}{2\pi} \int_0^{2\pi} f(\chi)\, d\chi$$

$$= -\frac{1}{(2\pi)^2} P \int_0^{2\pi} \cot\left(\frac{\chi - \theta}{2}\right) d\chi \cdot P \int_0^{2\pi} f(\chi') \cot\left(\frac{\chi' - \chi}{2}\right) d\chi' \qquad (60)$$

For the case of the half plane, we have likewise

$$g(x) = -\frac{1}{\pi} P \int_{-\infty}^{\infty} \frac{f(s)}{s - x}\, ds \qquad (61)$$

$$f(x) = \frac{1}{\pi} P \int_{-\infty}^{\infty} \frac{g(s)}{s - x}\, ds \qquad (62)$$

where f and g are the real and imaginary parts of the values $\Phi(x + i0)$ of an analytic function $\Phi(z) = \psi + i\phi$ over the boundary line $y = 0$. The function $g(x)$ is the *Hilbert transform* of $f(s)$. Denoting the integral operator

$$-\frac{1}{\pi} P \int_{-\infty}^{\infty} \frac{ds}{s - x} \qquad (63)$$

by $\boldsymbol{H}$ we obtain

$$\boldsymbol{H}\{\boldsymbol{H}[f]\} = -f \qquad (64)$$

Whereas the reciprocal formulae (61) and (62) involve integrations over the infinite range $-\infty < s < \infty$, the range of integration in the pair of formulae (58) and (59) is $0 \leqslant \chi \leqslant 2\pi$ and are consequently referred to as *finite Hilbert transforms*.

5.4   Singular integral equation of Hilbert type

Let us now consider the singular integral equation

$$a\phi(x) + \frac{b}{2\pi} P \int_0^{2\pi} \phi(s) \cot\left(\frac{s-x}{2}\right) ds = f(x) \tag{65}$$

having a singular kernel of the Hilbert type (43). Operating on both sides of (65) with

$$a\int_0^{2\pi} \delta(s-x)\, dx + \frac{b}{2\pi} P \int_0^{2\pi} \cot\left(\frac{s-x}{2}\right) dx$$

and setting

$$F(x) = af(x) - \frac{b}{2\pi} P \int_0^{2\pi} f(s) \cot\left(\frac{s-x}{2}\right) ds, \tag{66}$$

we obtain the simple result

$$(a^2 + b^2)\phi(x) - \frac{b^2}{2\pi} \int_0^{2\pi} \phi(s)\, ds = F(x) \tag{67}$$

on using the Hilbert formula (60).

To solve this equation we note that

$$(a^2 + b^2)\int_0^{2\pi} \phi(x)\, dx - b^2 \int_0^{2\pi} \phi(s)\, ds = \int_0^{2\pi} F(x)\, dx$$

and hence

$$\int_0^{2\pi} \phi(x)\, dx = \frac{1}{a^2}\int_0^{2\pi} F(x)\, dx = \frac{1}{a}\int_0^{2\pi} f(x)\, dx \tag{68}$$

since

$$\int_0^{2\pi} \cot\left(\frac{s-x}{2}\right) dx = 0.$$

Thus the solution of (65) is

$$\phi(x) = \frac{F(x)}{a^2 + b^2} + \frac{b^2}{2\pi a(a^2 + b^2)} \int_0^{2\pi} f(s)\, ds \tag{69}$$

If $a = 0$ (and we put $b = 1$) we obtain an integral equation of the first kind having the form

$$f(x) = \frac{1}{2\pi} P \int_0^{2\pi} \phi(s) \cot\left(\frac{s-x}{2}\right) ds. \tag{70}$$

Using (59) we arrive at the solution

$$\phi(x) = \frac{1}{2\pi} \int_0^{2\pi} \phi(s) \, ds - \frac{1}{2\pi} P \int_0^{2\pi} f(s) \cot\left(\frac{s-x}{2}\right) ds \tag{71}$$

If we now put

$$\frac{1}{2\pi} \int_0^{2\pi} \phi(s) \, ds = C$$

and substitute

$$\phi(x) = C - \frac{1}{2\pi} P \int_0^{2\pi} f(s) \cot\left(\frac{s-x}{2}\right) ds \tag{72}$$

back into (70), we see that (72) provides a solution for any value of the constant C if and only if

$$\int_0^{2\pi} f(x) \, dx = 0. \tag{73}$$

Finally we consider the singular integral equation of the second kind

$$\phi(x) - \frac{\lambda}{\pi} P \int_{-\infty}^{\infty} \frac{\phi(s)}{s-x} \, ds = f(x) \tag{74}$$

Operating on both sides of this equation with

$$\int_{-\infty}^{\infty} \delta(x-s) \, dx + \frac{\lambda}{\pi} P \int_{-\infty}^{\infty} \frac{dx}{x-s}$$

and setting

$$F(x) = f(x) + \frac{\lambda}{\pi} P \int_{-\infty}^{\infty} \frac{f(s)}{s-x} \, ds \tag{75}$$

we find that

$$(1 + \lambda^2)\phi(x) = F(x)$$

on using (64). Hence the solution of (74) is

$$\phi(x) = \frac{1}{1+\lambda^2}\left\{f(x) + \frac{\lambda}{\pi} P\int_{-\infty}^{\infty} \frac{f(s)}{s-x}\,ds\right\} \qquad (76)$$

Problems

1. By taking $f(2\pi-\chi)=f(\chi)$ and $g(2\pi-\chi)=-g(\chi)$ in the pair of finite Hilbert transforms (58) and (59), show that they may be rewritten in the form

$$(i) \quad g(\theta) = -\frac{1}{\pi} P\int_0^{\pi} f(\phi)\frac{\sin\theta}{\cos\theta-\cos\phi}\,d\phi$$

$$(ii) \quad f(\theta) = \frac{1}{\pi} P\int_0^{\pi} g(\phi)\frac{\sin\phi}{\cos\theta-\cos\phi}\,d\phi + \frac{1}{\pi}\int_0^{\pi} f(\phi)\,d\phi.$$

$$\left[\text{Hint: use } \tfrac{1}{2}\{\cot\tfrac{1}{2}(\theta+\phi)+\cot\tfrac{1}{2}(\theta-\phi)\} = \frac{\sin\theta}{\cos\phi-\cos\theta}\right]$$

2. By putting $x=\cos\theta$, $y=\cos\phi$ and

$$u(x) = f(\theta)/\sin\theta, \qquad v(x) = g(\theta)/\sin\theta$$

show that the pair of reciprocal formulae (i) and (ii) in problem 1 may be rewritten as

$$(i) \quad v(x) = -\frac{1}{\pi} P\int_{-1}^{1} \frac{u(y)}{x-y}\,dy$$

$$(ii) \quad u(x) = \frac{1}{\pi} P\int_{-1}^{1} \sqrt{\frac{1-y^2}{1-x^2}}\frac{v(y)}{x-y}\,dy + \frac{1}{\pi}\frac{1}{\sqrt{1-x^2}}\int_{-1}^{1} u(y)\,dy$$

3. Find the solution of

$$P\int_{-1}^{1} \frac{u(y)}{x-y}\,dy = 0$$

4. Find the solution of Föppl's integral equation

$$\frac{1}{\pi} P\int_{-1}^{1} \frac{\phi(t)}{t^2-s^2}\,dt = f(s)$$

where $\phi(t)$ and $f(s)$ are even functions.

5. Find the solution of

$$\frac{1}{\pi} P \int_{-a}^{a} \frac{tg(t)}{s^2 - t^2} \, dt = s^2.$$

6. Evaluate $\dfrac{1}{\pi} P \displaystyle\int_{-\infty}^{\infty} \dfrac{e^{is}}{s - x} \, ds$ using Cauchy's residue theorem and hence find the solutions of

$$\text{(i)} \quad \sin x = -\frac{1}{\pi} P \int_{-\infty}^{\infty} \frac{f(s)}{s - x} \, ds$$

$$\text{(ii)} \quad \cos x = \frac{1}{\pi} P \int_{-\infty}^{\infty} \frac{g(s)}{s - x} \, ds$$

Verify that (i) and (ii) form a pair of reciprocal Hilbert transforms.

7. Find the solution of

$$\frac{1}{1 + x^2} = P \frac{1}{\pi} \int_{-\infty}^{\infty} \frac{f(s)}{s - x} \, ds.$$

6

Hilbert space

In the remaining chapters of this book we shall be giving an introduction to the general theory of linear integral equations as developed by Volterra, Fredholm, Hilbert, and Schmidt. For this purpose we need to introduce the concept of a Hilbert space. This is a suitable generalization of ordinary three-dimensional Euclidean space to a linear vector space of infinite dimensions which, for the subject of integral equations, is chosen to be a complete linear space composed of square integrable functions having a distance property defined in terms of an inner product.

To explain the meanings of these terms we consider Euclidean space first and then the Hilbert space of sequences.

6.1   Euclidean space

In three-dimensional Euclidean space each point is specified by an ordered set of three real numbers or coordinates (x_1, x_2, x_3) forming the components of the position vector $\mathbf{x}$. The vector $\lambda\mathbf{x}$ has coordinates $(\lambda x_1, \lambda x_2, \lambda x_3)$ and the vector sum $\mathbf{x}+\mathbf{y}$ of two vectors $\mathbf{x}, \mathbf{y}$ has coordinates $(x_1+y_1, x_2+y_2, x_3+y_3)$. These are properties of a linear vector space.

The scalar product or inner product of two vectors $\mathbf{x}, \mathbf{y}$ is defined as

$$(\mathbf{x}, \mathbf{y}) = x_1 y_1 + x_2 y_2 + x_3 y_3 \tag{1}$$

and the vectors are orthogonal, i.e. at right angles, if $(\mathbf{x}, \mathbf{y}) = 0$.

We have

$$(\mathbf{x}, \mathbf{x}) = x_1^2 + x_2^2 + x_3^2 \geq 0,$$

where $(\mathbf{x}, \mathbf{x}) = 0$ if and only if $\mathbf{x}$ is the zero vector $\mathbf{0}$ with coordinates $(0, 0, 0)$.

The magnitude or norm $\|\mathbf{x}\|$ of a vector $\mathbf{x}$ is given by

$$\|\mathbf{x}\| = \sqrt{(\mathbf{x}, \mathbf{x})} = \sqrt{x_1^2 + x_2^2 + x_3^2} < \infty. \tag{2}$$

The vector is said to be normalized if $\|\mathbf{x}\| = 1$ and then $\mathbf{x}$ is a unit vector.

The distance between two points specified by the vectors $\mathbf{x}, \mathbf{y}$ is given by $\|\mathbf{x} - \mathbf{y}\|$. Evidently $\|\mathbf{x}\|$ is the distance of the point $\mathbf{x}$ from the origin specified by the zero vector $\mathbf{0}$.

Since the length of a side of a triangle is less than, or equal to when the triangle collapses into a straight line, the sum of the lengths of the other two sides, we have the triangle inequality

$$\|\mathbf{x} - \mathbf{y}\| \leqslant \|\mathbf{x}\| + \|\mathbf{y}\|. \tag{3}$$

Suppose that $\mathbf{x}$, $\mathbf{y}$, $\mathbf{z}$ are linearly independent vectors so that $\lambda\mathbf{x} + \mu\mathbf{y} + \nu\mathbf{z}$ is not the zero vector $\mathbf{0}$ except when $\lambda = \mu = \nu = 0$. Then we can use them to construct an orthogonal and normalized, i.e. orthonormal set of three vectors $\mathbf{e}_1$, $\mathbf{e}_2$, $\mathbf{e}_3$. Thus let us put $\mathbf{e}_1 = \mathbf{x}/\|\mathbf{x}\|$. This vector is clearly normalized. Then

$$\mathbf{y}' = \mathbf{y} - (\mathbf{y}, \mathbf{e}_1)\mathbf{e}_1$$

is orthogonal to $\mathbf{e}_1$ and $\mathbf{e}_2 = \mathbf{y}'/\|\mathbf{y}'\|$ is normalized.
Further

$$\mathbf{z}' = \mathbf{z} - (\mathbf{z}, \mathbf{e}_1)\mathbf{e}_1 - (\mathbf{z}, \mathbf{e}_2)\mathbf{e}_2$$

is orthogonal to $\mathbf{e}_1$ and $\mathbf{e}_2$ while $\mathbf{e}_3 = \mathbf{z}'/\|\mathbf{z}'\|$ is also normalized.

The three mutually orthogonal unit vectors $\mathbf{e}_1, \mathbf{e}_2, \mathbf{e}_3$ are said to form a *basis* since any vector $\mathbf{a}$ of the three-dimensional space can be expressed as the linear combination

$$\mathbf{a} = (\mathbf{a}, \mathbf{e}_1)\mathbf{e}_1 + (\mathbf{a}, \mathbf{e}_2)\mathbf{e}_2 + (\mathbf{a}, \mathbf{e}_3)\mathbf{e}_3.$$

The foregoing vector algebra can be extended to an n-dimensional space whose points are specified by an ordered set of n complex numbers $(x_1, x_2, \ldots, x_n)$ denoted by the vector $\mathbf{x}$. The inner product of two vectors $\mathbf{x}, \mathbf{y}$ is now defined as

$$(\mathbf{x}, \mathbf{y}) = \sum_{r=1}^{n} x_r \bar{y}_r = \overline{(\mathbf{y}, \mathbf{x})} \tag{4}$$

while the norm of the vector $\mathbf{x}$ is given by

$$\|\mathbf{x}\| = \sqrt{(\mathbf{x}, \mathbf{x})} = \sqrt{\sum_{r=1}^{n} |x_r|^2} < \infty. \tag{5}$$

An orthonormal set of n vectors $\mathbf{e}_1, \mathbf{e}_2, \ldots, \mathbf{e}_n$ form a basis of the n-dimensional space, and an arbitrary vector $\mathbf{a}$ in the space can be expressed as the linear combination

$$\mathbf{a} = \sum_{r=1}^{n} (\mathbf{a}, \mathbf{e}_r)\mathbf{e}_r$$

where the $a_r = (\mathbf{a}, \mathbf{e}_r)(r = 1, \ldots, n)$ are the components of the vector **a** with respect to the basis vectors. The only vector which is orthogonal to every vector of the basis is clearly the zero vector **0** and so the orthonormal set $\mathbf{e}_1, \mathbf{e}_2, \ldots, \mathbf{e}_n$ is said to *span* the n-dimensional space. If we take $\mathbf{e}_1 = (1, 0, 0, \ldots, 0)$, $\mathbf{e}_2 = (0, 1, 0, \ldots, 0), \ldots, \mathbf{e}_n = (0, 0, 0, \ldots, 1)$ we see that **a** is the vector specified by $(a_1, a_2, \ldots, a_n)$.

6.2   Hilbert space of sequences

By a natural generalization of a finite dimensional space, we can consider an infinite dimensional space whose points are represented by vectors **x** having components, or coordinates, given by the infinite sequence of complex numbers $\{x_r\} = (x_1, x_2, \ldots, x_r, \ldots)$ satisfying

$$\sum_{r=1}^{\infty} |x_r|^2 < \infty. \tag{6}$$

We now introduce the scalar product or *inner product* of two vectors **x** and **y** given by

$$(\mathbf{x}, \mathbf{y}) = \sum_{r=1}^{\infty} x_r \bar{y}_r = \overline{(\mathbf{y}, \mathbf{x})} \tag{7}$$

Then we have $0 \le (\mathbf{x}, \mathbf{x}) < \infty$, where $(\mathbf{x}, \mathbf{x}) = 0$ if and only if **x** is the zero vector **0** whose components all vanish.

Also we define the *norm* $\|\mathbf{x}\|$ of a vector **x** by the formula

$$\|\mathbf{x}\| = \sqrt{(\mathbf{x}, \mathbf{x})} = \sqrt{\sum_{r=1}^{\infty} |x_r|^2}. \tag{8}$$

Thus $\|\mathbf{x}\| = 0$ if and only if $\mathbf{x} = 0$.

Further we let $\lambda \mathbf{x}$ be the vector with components $\{\lambda x_r\}$ so that $\|\lambda \mathbf{x}\| = |\lambda| \|\mathbf{x}\|$, and let the sum $\mathbf{x} + \mathbf{y}$ of two vectors **x**, **y** be the vector having components $\{x_r + y_r\}$ as in the case of a finite dimensional space.

Now, by the Cauchy inequality

$$\sum_{r=1}^{\infty} |x_r| |y_r| \le \sqrt{\left(\sum_{r=1}^{\infty} |x_r|^2\right)\left(\sum_{r=1}^{\infty} |y_r|^2\right)} \tag{9}$$

and the inequality

$$\left| \sum_{r=1}^{\infty} x_r \bar{y}_r \right| \le \sum_{r=1}^{\infty} |x_r| \, |y_r|, \tag{10}$$

we obtain *Schwarz's inequality*

$$|(\mathbf{x}, \mathbf{y})| \le \|\mathbf{x}\| \, \|\mathbf{y}\|. \tag{11}$$

This is the generalization to infinite sequences of the corresponding result in three-dimensional Euclidean space which follows immediately from $(\mathbf{x}, \mathbf{y}) = \|\mathbf{x}\| \, \|\mathbf{y}\| \cos \alpha$ where α is the angle between the vectors $\mathbf{x}$ and $\mathbf{y}$.

Hence

$$\begin{aligned}
\|\mathbf{x} + \mathbf{y}\|^2 &= \sum_{r=1}^{\infty} |x_r + y_r|^2 \\
&= \sum_{r=1}^{\infty} |x_r|^2 + \sum_{r=1}^{\infty} |y_r|^2 + \sum_{r=1}^{\infty} (x_r \bar{y}_r + \bar{x}_r y_r) \\
&\le (\|\mathbf{x}\| + \|\mathbf{y}\|)^2 < \infty.
\end{aligned}$$

This shows that the sum $\mathbf{x} + \mathbf{y}$ satisfies the condition

$$\sum_{r=1}^{\infty} |x_r + y_r|^2 < \infty \tag{12}$$

and also yields the *triangle inequality*

$$\|\mathbf{x} + \mathbf{y}\| \le \|\mathbf{x}\| + \|\mathbf{y}\| \tag{13}$$

which can be rewritten in the form (3) by reversing the sign of $\mathbf{y}$.

The real number

$$d(\mathbf{x}, \mathbf{y}) = \|\mathbf{x} - \mathbf{y}\| = \sqrt{\sum_{r=1}^{\infty} |x_r - y_r|^2} \tag{14}$$

represents the *distance* between two points characterized by vectors $\mathbf{x}$, $\mathbf{y}$ and is an obvious generalization of the idea of distance in three-dimensional Euclidean space. Clearly $\|\mathbf{x}\|$ is the distance of the point $\mathbf{x}$ from the origin given by the zero vector $\mathbf{0}$.

A sequence of vectors $\{\mathbf{x}_n\}$ *converges strongly* to a limit vector $\mathbf{x}$ if, given any $\varepsilon > 0$, there exists N such that for $n > N$ we have $\|\mathbf{x}_n - \mathbf{x}\| < \varepsilon$. Strong convergence is denoted by $\mathbf{x}_n \to \mathbf{x}$.

If $\mathbf{x}_n \to \mathbf{x}$ we have, using the triangle inequality,

$$\begin{aligned}
\|\mathbf{x}_n - \mathbf{x}_m\| &= \|\mathbf{x}_n - \mathbf{x} + \mathbf{x} - \mathbf{x}_m\| \\
&\le \|\mathbf{x}_n - \mathbf{x}\| + \|\mathbf{x}_m - \mathbf{x}\| < \varepsilon
\end{aligned}$$

for sufficiently large n and m. A sequence $\{\mathbf{x}_n\}$ satisfying $\|\mathbf{x}_n - \mathbf{x}_m\| < \varepsilon$ for sufficiently large n, m is known as a *Cauchy sequence*.

We shall now demonstrate the converse of the above result, namely that every Cauchy sequence has a limit vector $\mathbf{x}$ in the space.

Suppose that $\mathbf{e}_1 = (1, 0, 0, \ldots)$, $\mathbf{e}_2 = (0, 1, 0, \ldots), \ldots$ are unit vectors in the infinite dimensional space. They form a *basis* which spans the space, and an arbitrary vector $\mathbf{a} = (a_1, a_2, \ldots, a_r, \ldots)$ can be expressed as

$$\mathbf{a} = \sum_{r=1}^{\infty} a_r \mathbf{e}_r$$

where $a_r = (\mathbf{a}, \mathbf{e}_r)$ and $\mathbf{e}_r$ is the rth unit vector satisfying $\|\mathbf{e}_r\| = 1$.

Then we have, using Schwarz's inequality,

$$|(\mathbf{x}_n, \mathbf{e}_r) - (\mathbf{x}_m, \mathbf{e}_r)| = |(\mathbf{x}_n - \mathbf{x}_m, \mathbf{e}_r)|$$
$$\leqslant \|\mathbf{x}_n - \mathbf{x}_m\| < \varepsilon$$

for all sufficiently large n and m. It follows that the sequence of numbers $(\mathbf{x}_n, \mathbf{e}_r) = x_r^{(n)}$ is a Cauchy sequence and approaches a limiting value $x_r (r = 1, 2, \ldots)$ as $n \to \infty$. But for sufficiently large n, m we have

$$\|\mathbf{x}_n - \mathbf{x}_m\| = \sqrt{\sum_{r=1}^{\infty} |x_r^{(n)} - x_r^{(m)}|^2} < \varepsilon$$

and so for every k

$$\sqrt{\sum_{r=1}^{k} |x_r^{(n)} - x_r^{(m)}|^2} < \varepsilon.$$

Hence, in the limit as $m \to \infty$ we obtain

$$\sqrt{\sum_{r=1}^{k} |x_r^{(n)} - x_r|^2} < \varepsilon$$

and since this is true for every k we get

$$\|\mathbf{x}_n - \mathbf{x}\| = \sqrt{\sum_{r=1}^{\infty} |x_r^{(n)} - x_r|^2} < \varepsilon.$$

But

$$\sqrt{\sum_{r=1}^{\infty} |x_r|^2} = \|\mathbf{x}\| = \|(\mathbf{x} - \mathbf{x}_n) + \mathbf{x}_n\|$$
$$\leqslant \|\mathbf{x} - \mathbf{x}_n\| + \|\mathbf{x}_n\|$$
$$< \varepsilon + \|\mathbf{x}_n\|$$

and so

$$\sum_{r=1}^{\infty} |x_r|^2 < \infty.$$

Thus $\mathbf{x} = (x_1, x_2, \ldots, x_r, \ldots)$ belongs to the space and $\mathbf{x}_n \to \mathbf{x}$ as $n \to \infty$.

A space in which $\mathbf{x}_n \to \mathbf{x}$ when $\{\mathbf{x}_n\}$ is a Cauchy sequence is called *complete*.

The space of sequences described above is an example of a Hilbert space. We denote this space by l_2.

6.3   Function space

We consider the function space composed of all sectionally continuous complex functions $f(x)$ of a real variable x, defined in the interval $a \le x \le b$, which are square integrable and thus satisfy the condition

$$\int_a^b |f(x)|^2 \, dx < \infty. \tag{15}$$

Introducing the *inner product* of two such functions $f(x)$ and $g(x)$ given by

$$(f, g) = \int_a^b f(x)\overline{g(x)} \, dx, \tag{16}$$

we define the *norm* of the function $f(x)$ as

$$\|f\| = \sqrt{(f, f)}. \tag{17}$$

Next we establish the important inequality named after Schwarz. We have

$$\int_a^b \left| f(x) - \frac{(f, g)}{(g, g)} g(x) \right|^2 dx \ge 0 \tag{18}$$

so that

$$(f, f) - 2\frac{|(f, g)|^2}{(g, g)} + \frac{|(f, g)|^2}{(g, g)} \ge 0$$

i.e.

$$(f, f)(g, g) \ge |(f, g)|^2 \tag{19}$$

Hence

$$\|f\|\,\|g\| \geqslant |(f, g)| \qquad (20)$$

which is *Schwarz's inequality* for square integrable functions.

Also

$$(\|f\| + \|g\|)^2 = \|f\|^2 + \|g\|^2 + 2\|f\|\,\|g\|$$
$$\geqslant (f, f) + (g, g) + 2|(f, g)|$$

by Schwarz's inequality. But

$$2\,|(f, g)| \geqslant (f, g) + \overline{(f, g)}$$

and so

$$(\|f\| + \|g\|)^2 \geqslant (f, f) + (g, g) + (f, g) + (g, f)$$
$$= (f + g, f + g).$$

Hence we have

$$\|f\| + \|g\| \geqslant \|f + g\| \qquad (21)$$

which is the *triangle inequality* for functions, sometimes known as *Minkowski's inequality*.

6.3.1 Orthonormal system of functions

Two functions $f(x)$ and $g(x)$ belonging to the function space are said to be *orthogonal* if

$$(f, g) = \int_a^b f(x)\overline{g(x)}\,\mathrm{d}x = 0 \qquad (22)$$

and the function $f(x)$ is normalized if

$$\|f\| = 1. \qquad (23)$$

We consider a set of sectionally continuous complex functions $\phi_1(x), \phi_2(x), \ldots, \phi_r(x), \ldots$ satisfying the orthonormality condition

$$(\phi_r, \phi_s) = \int_a^b \phi_r(x)\overline{\phi_s(x)}\,\mathrm{d}x = \delta_{rs} \qquad (24)$$

where δ_{rs} is the Kronecker delta symbol

$$\delta_{rs} = \begin{cases} 1 & (r = s) \\ 0 & (r \neq s). \end{cases} \qquad (25)$$

Such a set of functions is called *orthonormal*.

An orthonormal system of functions is said to form a *basis* or a *complete system** if and only if the sole function which is orthogonal to every member $\phi_r(x)$ of the system is the null function which vanishes throughout the interval $a \leq x \leq b$ except at a finite number of points.

It can be shown that every orthonormal system is either finite or denumerably infinite, i.e., the members of the system can be placed in 1-1 correspondence with the natural numbers, and that an incomplete system can always be completed to form a basis by adding a finite or denumerable set of functions.

6.3.2   Gram-Schmidt orthogonalization

Now suppose that we have a finite or denumerable system of functions $\chi_1(x), \chi_2(x), \ldots, \chi_r(x), \ldots$ which is not orthonormal. We assume that the functions are linearly independent and that none of them is null. Then if $\lambda_1\chi_1 + \lambda_2\chi_2 + \ldots + \lambda_n\chi_n$ is the null function we must have $\lambda_1 = \lambda_2 = \ldots = \lambda_n = 0$ and this holds for all n. We aim to construct out of this system of functions, a new system which is orthonormal, just as we were able to do in section 6.1 for three-dimensional Euclidean space. We proceed by using the principle of induction.

Clearly

$$\phi_1(x) = \frac{\chi_1(x)}{\|\chi_1\|} \tag{26}$$

is normalized. Further

$$\phi_2(x) = \frac{\psi_2(x)}{\|\psi_2\|}, \tag{27}$$

where

$$\psi_2(x) = \chi_2(x) - (\chi_2, \phi_1)\phi_1(x), \tag{28}$$

is normalized and also orthogonal to $\phi_1(x)$.

Now let us suppose that we have constructed n such functions $\phi_1(x), \phi_2(x), \ldots, \phi_n(x)$ which are normalized and mutually orthogonal. Then

$$\phi_{n+1}(x) = \frac{\psi_{n+1}(x)}{\|\psi_{n+1}\|}, \tag{29}$$

---

* A complete system of functions should not be confused with a complete space, discussed in section 6.3.4.

where

$$\psi_{n+1}(x) = \chi_{n+1}(x) - \sum_{r=1}^{n} (\chi_{n+1}, \phi_r) \phi_r(x), \qquad (30)$$

is normalized and orthogonal to all the functions $\phi_1(x), \ldots, \phi_n(x)$. Thus, using the principle of induction, we have shown that it is possible to construct an orthonormal set of functions from the original set.

The above process is called Gram-Schmidt orthogonalization.

6.3.3 Mean square convergence

Let $f(x)$ be any function belonging to the function space and suppose that we wish to obtain a best approximation to $f(x)$ in the form of the sum

$$\sum_{r=1}^{n} c_r \phi_r(x) \qquad (31)$$

where the c_r are parameters to be determined.

We can achieve this by requiring that

$$I_n = \int_a^b \left| f(x) - \sum_{r=1}^{n} c_r \phi_r(x) \right|^2 \mathrm{d}x \qquad (32)$$

be chosen as small as possible. This will then provide us with the *best mean square approximation* to $f(x)$.

We have that

$$I_n = (f, f) - \sum_{r=1}^{n} \{\bar{c}_r(f, \phi_r) + c_r \overline{(f, \phi_r)}\} + \sum_{r=1}^{n} |c_r|^2$$

$$= (f, f) + \sum_{r=1}^{n} |c_r - a_r|^2 - \sum_{r=1}^{n} |a_r|^2 \qquad (33)$$

where the $a_r = (f, \phi_r)$ are known as the *Fourier coefficients* of $f(x)$. Evidently I_n attains its least value when $c_r = a_r (r = 1, \ldots, n)$ and then we have

$$I_n = (f, f) - \sum_{r=1}^{n} |a_r|^2. \qquad (34)$$

Since $I_n \geq 0$ it follows at once that

$$(f, f) \geq \sum_{r=1}^{n} |a_r|^2 \qquad (35)$$

for all n. This is called *Bessel's inequality*.

If now

$$\lim_{n \to \infty} I_n = 0 \tag{36}$$

we say that $f(x)$ is the *mean square limit* of the sequence of functions $\{f_n(x)\}$ given by

$$f_n(x) = \sum_{r=1}^{n} a_r \phi_r(x), \tag{37}$$

which we may write $\|f - f_n\| \to 0$ as $n \to \infty$, or that $\{f_n(x)\}$ is *strongly convergent* to $f(x)$, written $f_n(x) \to f(x)$ as $n \to \infty$.

Then it follows that

$$\|f\|^2 = \sum_{r=1}^{\infty} |(f, \phi_r)|^2 \tag{38}$$

which is known as Parseval's formula or the *completeness relation*. If this formula holds, the orthonormal system of functions $\phi_r(x)$ forms a basis or a complete system. For suppose that, on the contrary, there exists a function $f(x)$ with non-vanishing norm in the function space satisfying $(f, \phi_r) = 0$ for all values of r. Then it follows from Parseval's formula that $\|f\| = 0$ which provides a contradiction.

Now consider two functions $f(x)$ and $g(x)$ belonging to the function space. Then

$$\|f + \lambda g\|^2 = \sum_{r=1}^{\infty} |(f + \lambda g, \phi_r)|^2$$

supposing that the system of functions ϕ_r forms a basis, and so

$$\|f\|^2 + |\lambda|^2 \|g\|^2 + \bar{\lambda}(f, g) + \lambda(g, f)$$
$$= \sum_{r=1}^{\infty} |(f, \phi_r)|^2 + |\lambda|^2 \sum_{r=1}^{\infty} |(g, \phi_r)|^2$$
$$+ \sum_{r=1}^{\infty} \{\bar{\lambda}(f, \phi_r)(\phi_r, g) + \lambda(g, \phi_r)(\phi_r, f)\}.$$

Since λ is an arbitrary parameter we see that

$$(f, g) = \sum_{r=1}^{\infty} (f, \phi_r)(\phi_r, g) \tag{39}$$

which is known as the *generalized Parseval formula*.

6.3.4   Riesz-Fischer theorem

A sequence of functions $\{f_n(x)\}$ is called a *Cauchy sequence* if

$$\lim_{n,m\to\infty} \|f_n - f_m\| = 0. \tag{40}$$

If a sequence $\{f_n(x)\}$ is mean square convergent to $f(x)$ it is a Cauchy sequence, for using the triangle inequality we see that

$$\|f_m - f_n\| = \|(f_m - f) + (f - f_n)\|$$
$$\leqslant \|f_m - f\| + \|f - f_n\|$$
$$\to 0 \qquad \text{as } m, n \to \infty.$$

A function space is said to be *complete* if every Cauchy sequence $\{f_n(x)\}$ is mean square convergent to a function $f(x)$ belonging to the space, i.e., $f_n(x)$ is strongly convergent to $f(x)$ or $f_n(x) \to f(x)$.

Now, so far, we have assumed that our function space is composed of sectionally continuous functions and such a space is *not* complete. However a function space composed of functions that are square integrable in the Lebesgue sense, i.e., L^2 functions, is complete. This result is Fischer's form of the Riesz-Fischer theorem which we assert without proof. It leads directly to Riesz's form of the Riesz-Fischer theorem:

If $\phi_1(x), \phi_2(x), \dots, \phi_r(x), \dots$ is an orthonormal system of L^2 functions and $\{a_r\}$ is a sequence of complex numbers, then

$$\sum_{r=1}^{n} a_r \phi_r(x)$$

is mean square convergent to a L^2 function $f(x)$ whose Fourier coefficients are $\{a_r\}$ if and only if

$$\sum_{r=1}^{\infty} |a_r|^2 < \infty.$$

The space of L^2 functions $f(x)$ defined over the interval $a \leqslant x \leqslant b$ is another example of a Hilbert space, usually denoted by L_2.

In this space $f = g$ if the functions f and g are equal 'almost everywhere', that is everywhere except at a finite number or a denumerable infinity of points of the interval $a \leqslant x \leqslant b$. Further $f = 0$ in this space if f vanishes 'almost everywhere'.

6.4   Abstract Hilbert space H

We have described above two examples of Hilbert spaces, the space of sequences l_2 and the space of L^2 functions. We shall conclude this chapter by abstracting the common axioms that all Hilbert spaces must satisfy.

A Hilbert space H is a *complete linear vector space* possessing a *distance function* or *metric* which is given by an *inner product*.

A *linear vector space* is a set of *elements*, sometimes called points or *vectors*, f, g, $h, \ldots$ forming an *Abelian group* and permitting multiplication by the *field* of complex numbers λ.

An Abelian group has an *internal law of composition* denoted by the addition sign $+$ satisfying the *commutative law*

$$f + g = g + f \tag{41}$$

and the *associative law*

$$f + (g + h) = (f + g) + h, \tag{42}$$

having a zero element 0 such that

$$0 + f = f + 0 = f, \tag{43}$$

and an *inverse* element $-f$ corresponding to each element f of the set such that

$$f + (-f) = (-f) + f = 0. \tag{44}$$

The multiplication by the field of complex numbers satisfies

$$1 \cdot f = f, \tag{45}$$

$$0 \cdot f = 0, \tag{46}$$

$$(\lambda\mu)f = \lambda(\mu f), \tag{47}$$

and satisfies the *distributive law* with respect to the elements f, g

$$\lambda(f + g) = \lambda f + \lambda g \tag{48}$$

and the distributive law with respect to the numbers λ, μ

$$(\lambda + \mu)f = \lambda f + \mu f. \tag{49}$$

The *inner product* of two elements f, g is a complex number denoted by (f, g) satisfying the conditions

$$(f, g) = \overline{(g, f)}, \tag{50}$$

$$(\lambda f, g) = \lambda(f, g), \tag{51}$$

$$(f_1 + f_2, g) = (f_1, g) + (f_2, g). \tag{52}$$

It follows at once that $(f, f) = \overline{(f, f)}$ so that (f, f) is a real number. We shall assume that

$$(f, f) \geqslant 0 \tag{53}$$

and further that $(f, f) = 0$ if and only if $f = 0$.

Then we have also that

$$(f, \lambda g) = \overline{(\lambda g, f)} = \bar{\lambda}\overline{(g, f)} = \bar{\lambda}(f, g) \tag{54}$$

and

$$\begin{aligned}(f, g_1 + g_2) &= \overline{(g_1 + g_2, f)} = \overline{(g_1, f)} + \overline{(g_2, f)} \\ &= (f, g_1) + (f, g_2).\end{aligned} \tag{55}$$

The norm of the element f is defined as

$$\|f\| = \sqrt{(f, f)}. \tag{56}$$

We see that $\|f\| = 0$ if and only if $f = 0$, and $\|\lambda f\| = |\lambda|\,\|f\|$.

Now

$$(f + \lambda g, f + \lambda g) \geqslant 0$$

and so taking $\lambda = -(f, g)/(g, g)$ we obtain Schwarz's inequality

$$\|f\|\,\|g\| \geqslant |(f, g)|. \tag{57}$$

Again following the proof given in section 6.3 we arrive at the triangle inequality

$$\|f\| + \|g\| \geqslant \|f + g\|. \tag{58}$$

We now define a *distance function* $d(f, g)$ in terms of the norm according to the formula

$$d(f, g) = \|f - g\|. \tag{59}$$

This satisfies the conditions required of a distance between two points f and g, namely

$$\left.\begin{array}{ll}\text{(i)} & d(f, g) = d(g, f), \\ \text{(ii)} & d(f, g) \geqslant 0, \\ \text{(iii)} & d(f, g) = 0 \text{ if and only if } f = g, \\ \text{(iv)} & d(f, g) \leqslant d(f, h) + d(h, g).\end{array}\right\} \tag{60}$$

This last condition follows from the triangle inequality:

$$\|f - g\| = \|(f - h) + (h - g)\| \leqslant \|f - h\| + \|h - g\|.$$

A sequence of elements $\{f_n\}$ *converges strongly* to a *limit element f* if, given any $\varepsilon > 0$, there exists a N such that for $n > N$ we have $\|f_n - f\| < \varepsilon$. Strong convergence is denoted by $f_n \to f$.

If $f_n \to f$ we have

$$\|f_n - f_m\| = \|(f_n - f) + (f - f_m)\|$$
$$\leqslant \|f_n - f\| + \|f_m - f\| < \varepsilon$$

for sufficiently large n, m. A sequence $\{f_n\}$ of elements satisfying $\|f_n - f_m\| < \varepsilon$ for sufficiently large n, m is known as a *Cauchy sequence*.

A Hilbert space is *complete*, that is every Cauchy sequence converges to a limit vector in the space.

6.4.1 Dimension of Hilbert space

To define the *dimension* of a space R we introduce the following concepts. A subset T of R is said to be *dense* in R if, given any $\varepsilon > 0$ and any element $f \in R$, there exists an element $g \in T$ such that $\|f - g\| < \varepsilon$. S is called a *fundamental set* in R if the set T of all linear combinations $\sum_{r=1}^{\infty} c_r f_r$ with $f_r \in S$ is dense in R. Then the dimension of the space R is the least possible cardinal number of a fundamental set S in R.

The dimension of Hilbert space H is required to be denumerably infinite.

6.4.2 Complete orthonormal system

Any set of linearly independent elements belonging to H can be combined together to form an orthonormal system $\phi_1, \phi_2, \ldots, \phi_r, \ldots$ satisfying

$$(\phi_r, \phi_s) = \delta_{rs}$$

by applying the Gram-Schmidt orthogonalization process described in section 6.3.2.

Because the dimension of H is denumerably infinite it follows that any orthonormal system of elements belonging to H has not more than a denumerable infinity of elements which we may denote by $\phi_1, \phi_2, \ldots, \phi_r, \ldots$

Given any element $f \in H$ and putting

$$f_n = \sum_{r=1}^{n} (f, \phi_r) \phi_r \tag{61}$$

where the (f, ϕ_r) are the Fourier coefficients of f, we can show that

$$\|f - f_n\|^2 = \|f\|^2 - \sum_{r=1}^{n} |(f, \phi_r)|^2 \tag{62}$$

by following the analysis given in section 6.3.3. Since $\|f - f_n\| \geq 0$, we obtain Bessel's inequality

$$\|f\|^2 \geq \sum_{r=1}^{n} |(f, \phi_r)|^2, \tag{63}$$

valid for all n, from which it follows that the series on the right-side converges as $n \to \infty$.

Hence, given any $\varepsilon > 0$, we have

$$\|f_n - f_m\|^2 = \left\| \sum_{r=m+1}^{n} (f, \phi_r)\phi_r \right\|^2 \quad (n > m)$$

$$= \sum_{r=m+1}^{n} |(f, \phi_r)|^2 < \varepsilon$$

provided n, m are sufficiently large. Therefore $\{f_n\}$ is a Cauchy sequence and so there exists an element $f' \in H$ such that $f_n \to f'$ as $n \to \infty$, using the fact that H is complete. Thus

$$f' = \sum_{r=1}^{\infty} (f, \phi_r)\phi_r \tag{64}$$

and

$$\|f\|^2 \geq \|f'\|^2. \tag{65}$$

If the system $\phi_1, \phi_2, \ldots, \phi_r, \ldots$ is a fundamental set in H then $f' = f$ and we have

$$\|f\|^2 = \sum_{r=1}^{\infty} |(f, \phi_r)|^2. \tag{66}$$

Then $\phi_1, \phi_2, \ldots, \phi_r, \ldots$ is called a *complete system* or *basis* and (66) is known as Parseval's formula or the *completeness relation*.

Following the analysis at the end of section 6.3.3 we can also show that the generalized Parseval formula (39) holds for $f, g \in H$.

Problems

1. Construct an orthonormal set of functions defined in the interval $-1 \leq x \leq 1$ from $1, x, x^2, x^3$ using the Gram-Schmidt orthogonalization process. Verify that the functions are proportional to the first four Legendre polynomials.

2. Show that the set of orthogonal functions $\{\cos nx\}$ $(n = 0, 1, 2, \ldots)$ do not form a basis spanning the space of continuous functions defined over the interval $-\pi \leq x \leq \pi$. Likewise show that the set of

orthogonal functions $\{\sin nx\}$ $(n = 1, 2, \ldots)$ do not form a basis of the space.

3. Show that the sequence of functions $\{f_n(x)\}$ where

$$f_n(x) = \begin{cases} 0 & \left(0 \leq x \leq \dfrac{1}{n}\right) \\[2mm] \sqrt{n} & \left(\dfrac{1}{n} < x < \dfrac{2}{n}\right) \\[2mm] 0 & \left(\dfrac{2}{n} \leq x \leq 1\right), \end{cases}$$

converges to zero at all points of the interval $0 \leq x \leq 1$ but that the sequence is not mean square convergent to zero.

4. Establish that the space of continuous functions is not complete by showing that the sequence of continuous functions $\{f_n(x)\}$ where

$$f_n(x) = \begin{cases} 0 & \left(-1 \leq x \leq -\dfrac{1}{n}\right) \\[2mm] \tfrac{1}{2}(nx+1) & \left(-\dfrac{1}{n} < x < \dfrac{1}{n}\right) \\[2mm] 1 & \left(\dfrac{1}{n} \leq x \leq 1\right), \end{cases}$$

converges strongly to the discontinuous function

$$f(x) = \begin{cases} 0 & (-1 \leq x < 0) \\[1mm] \tfrac{1}{2} & (x = 0) \\[1mm] 1 & (0 < x \leq 1). \end{cases}$$

5. Show that if a Cauchy sequence $\{f_n\}$ converges to an element f then f is unique.

6. Verify that the space of infinite sequences l_2 discussed in section 6.2 satisfies the axioms for abstract Hilbert space.

7. By considering the Cauchy sequence $\{\mathbf{x}_n\}$ where
$$\mathbf{x}_n = (1, \tfrac{1}{2}, \tfrac{1}{3}, \ldots, 1/n, 0, 0, \ldots),$$
show that the linear vector space of infinite sequences
$$(x_1, x_2, \ldots, x_r, \ldots)$$
in which only a finite number of the coordinates x_r do not vanish, is *not* complete.

Linear operators in Hilbert space

In an integral equation the unknown function occurs under the integral sign and thus, if the functions involved belong to a Hilbert space, it is clear that we have to deal with integral operators acting on a Hilbert space of functions.

We pointed out in section 6.3.4 that square integrable functions in the Lebesgue sense, that is L^2 functions, form a Hilbert space and consequently the appropriate integral operators have L^2 kernels introduced in section 1.4. However, in order to avoid unduly difficult concepts, we shall suppose that our functions and kernels are square integrable without usually specifying the sense in which the integrals are to be performed.

It is worth while placing linear integral operators in a more general context and so we shall conclude this chapter by giving an introduction to the theory of linear operators in an abstract Hilbert space.

7.1 Linear integral operators

We consider the linear integral operator

$$\boldsymbol{K} = \int_a^b K(x, s) \, \mathrm{d}s \tag{1}$$

where $K(x, s)$ is a square integrable kernel, and write

$$\psi(x) = \int_a^b K(x, s)\phi(s) \, \mathrm{d}s, \tag{2}$$

where $\phi(s)$ is a square integrable function, in the symbolic form

$$\psi = \boldsymbol{K}\phi. \tag{3}$$

It is evident that the operator $\boldsymbol{K}$ is linear since

$$\boldsymbol{K}(\lambda_1\phi_1 + \lambda_2\phi_2) = \lambda_1\boldsymbol{K}\phi_1 + \lambda_2\boldsymbol{K}\phi_2 \tag{4}$$

where λ_1, λ_2 are constants and ϕ_1, ϕ_2 are square integrable functions.

We also introduce the identity operator I satisfying

$$I\phi = \phi \tag{5}$$

for every square integrable function $\phi(s)$, which may be expressed in the form of the integral operator

$$I = \int_a^b \delta(x - s)\, \mathrm{d}s \tag{6}$$

where δ is the Dirac delta function defined in section 2.2.

If

$$L = \int_a^b L(x, s)\, \mathrm{d}s \tag{7}$$

is a second integral operator we have

$$\chi = L\psi = L(K\phi) \tag{8}$$

where

$$\chi(x) = \int_a^b L(x, t)\, \mathrm{d}t \int_a^b K(t, s)\phi(s)\, \mathrm{d}s$$

$$= \int_a^b P(x, s)\phi(s)\, \mathrm{d}s \tag{9}$$

and

$$P(x, s) = \int_a^b L(x, t)K(t, s)\, \mathrm{d}t, \tag{10}$$

that is

$$\chi = P\phi \tag{11}$$

where $P = LK$ is the integral operator with kernel $P(x, s)$.

Integral operators satisfy the associative law

$$M(LK) = (ML)K \tag{12}$$

and the distributive laws

$$M(L + K) = ML + MK \tag{13}$$

$$(L + K)M = LM + KM \tag{14}$$

but, in general, do not satisfy the commutative law.

Using the associative law we see that

$$\boldsymbol{K}^m \boldsymbol{K}^n = \boldsymbol{K}^{m+n}, \qquad (\boldsymbol{K}^m)^n = \boldsymbol{K}^{mn} \qquad (m, n \geq 1) \qquad (15)$$

where

$$\boldsymbol{K}^n = \int_a^b K_n(x, s)\, \mathrm{d}s \qquad (16)$$

and $K_n(x, s)$ is the iterated kernel defined by (4.22).

7.1.1 Norm of an integral operator

If $K(x, s)$ is a square integrable kernel its norm is defined by

$$\|\boldsymbol{K}\|_2 = \left[\int_a^b \int_a^b |K(x, s)|^2\, \mathrm{d}x\, \mathrm{d}s \right]^{1/2}. \qquad (17)$$

Then if $\phi(s)$ is a square integrable function and $\psi(x)$ is given by (2) we have, using Schwarz's inequality (6.20), that

$$|\psi(x)|^2 \leq \int_a^b |K(x, s)|^2\, \mathrm{d}s \int_a^b |\phi(s)|^2\, \mathrm{d}s$$

which yields

$$\int_a^b |\psi(x)|^2\, \mathrm{d}x \leq \int_a^b \int_a^b |K(x, s)|^2\, \mathrm{d}x\, \mathrm{d}s \int_a^b |\phi(s)|^2\, \mathrm{d}s$$

so that

$$\|\psi\| \leq \|\boldsymbol{K}\|_2 \|\phi\| < \infty \qquad (18)$$

and thus $\psi(x)$ is square integrable.

When $\|\boldsymbol{K}\|_2 = 0$ then $\|\psi\| = 0$ so that $\boldsymbol{K}\phi$ vanishes 'almost everywhere' for all square integrable functions ϕ, and $\boldsymbol{K}$ is called a *null operator*.

Also if $L(x, t)$ and $K(t, s)$ are square integrable kernels, then $P(x, s)$ given by (10) is square integrable since, by Schwarz's inequality,

$$|P(x, s)|^2 \leq \int_a^b |L(x, t)|^2\, \mathrm{d}t \int_a^b |K(t, s)|^2\, \mathrm{d}t$$

so that

$$\int_a^b \int_a^b |P(x, s)|^2 \, \mathrm{d}x \, \mathrm{d}s \leq \int_a^b \int_a^b |L(x, t)|^2 \, \mathrm{d}x \, \mathrm{d}t \int_a^b \int_a^b |K(t, s)|^2 \, \mathrm{d}t \, \mathrm{d}s$$

giving

$$\|LK\|_2 \leq \|L\|_2 \|K\|_2 < \infty. \tag{19}$$

Putting $L = K$ we see that $\|K^2\|_2 \leq \|K\|_2^2$ and in general

$$\|K^n\|_2 \leq \|K\|_2^n \qquad (n = 1, 2, 3, \ldots) \tag{20}$$

7.1.2 Hermitian adjoint

The Hermitian adjoint of a kernel $K(x, s)$ is defined to be the kernel

$$K^*(x, s) = \overline{K(s, x)}. \tag{21}$$

We see at once that

$$(\lambda K)^* = \bar{\lambda} K^*, \qquad K^{**} = K. \tag{22}$$

Also

$$(LK)^*(x, s) = \overline{LK(s, x)}$$

$$= \int_a^b \overline{L(s, t)} \, \overline{K(t, x)} \, \mathrm{d}t$$

$$= \int_a^b K^*(x, t) L^*(t, s) \, \mathrm{d}t$$

$$= K^* L^*(x, s)$$

and so

$$(LK)^* = K^* L^*. \tag{23}$$

Further, using the definition of the inner product given by (6.16), we have

$$(K\phi, \psi) = \int_a^b \overline{\psi(x)} \, \mathrm{d}x \int_a^b K(x, s) \phi(s) \, \mathrm{d}s$$

$$= \int_a^b \phi(s) \, \mathrm{d}s \int_a^b K(x, s) \overline{\psi(x)} \, \mathrm{d}x$$

$$= (\phi, K^* \psi). \tag{24}$$

If $K^* = K$ the kernel is called *Hermitian* or *self adjoint*.

7.2   Bounded linear operators

So far we have been concerned with linear integral operators acting on a space of square integrable functions which, if chosen to be all the functions which are integrable in the Lebesgue sense, would form a Hilbert space of functions L_2.

We shall now turn our attention to the case of an abstract Hilbert space, defined in section 6.4, acted on by bounded linear operators.

Consider a subset D of an abstract Hilbert space H such that if $f, g \in D$ then $\lambda f + \mu g \in D$ where λ, μ are arbitrary complex numbers. Then D is called a *linear manifold*. We note that a linear manifold must contain the zero element since $f + (-1)f = 0$.

Suppose that corresponding to any element $f \in D$ we assign an element $Kf \in \Delta$ where Δ is also a linear manifold. Then K maps D onto Δ and is called a *linear operator* with *domain D* and *range Δ* if

$$K(\lambda f + \mu g) = \lambda K f + \mu K g \tag{25}$$

for $f, g \in D$.

A linear operator K having the Hilbert space H as domain is *bounded* if there exists a constant $C \geqslant 0$ such that

$$\|Kf\| \leqslant C \|f\|$$

for all $f \in H$.

The *norm* $\|K\|$ of the bounded linear operator K is defined as the smallest possible value of C. Thus $\|K\|$ is the least upper bound or supremum of $\|Kf\|/\|f\|$, that is

$$\|K\| = \sup_{f \in H} \frac{\|Kf\|}{\|f\|} \tag{26}$$

and so

$$\|Kf\| \leqslant \|K\| \|f\|. \tag{27}$$

This is a generalization of the inequality (18) for the linear integral operator (1) with square integrable kernel. Clearly (1) is a bounded linear operator.

The linear operator K has an *adjoint operator* K^* defined so that

$$(Kf, g) = (f, K^*g) \tag{28}$$

for all $f, g \in H$. This is in accordance with the definition (21) for linear integral operators, as shown in section 7.1.2. If $K^* = K$ the operator is self adjoint or *Hermitian*.

Now by Schwarz's inequality (6.57), and (27), we have

$$|(Kf, g)| \leqslant \|Kf\| \|g\| \leqslant \|K\| \|f\| \|g\| \tag{29}$$

and so (Kf, g) is bounded. Likewise

$$|(f, K^*g)| \leqslant \|K^*\| \|f\| \|g\| \tag{30}$$

so that

$$\|K^*\| = \|K\| \tag{31}$$

and K^* is also a bounded linear operator.

Further, any bounded operator K is a *continuous* linear operator which transforms a strongly convergent sequence $\{f_n\}$ into a strongly convergent sequence $\{Kf_n\}$. For we have

$$\|Kf_n - Kf\| = \|K(f_n - f)\| \leqslant \|K\| \|f_n - f\|$$

so that if $f_n \to f$ as $n \to \infty$ then

$$\|Kf_n - Kf\| \to 0,$$

that is Kf_n is strongly convergent to Kf.

Actually every continuous linear operator K with domain H is bounded. For otherwise there would exist a sequence $\{f_n\}$ such that $\|Kf_n\| \geqslant n\|f_n\|$ so that $\|Kg_n\| \geqslant 1$, but $g_n = f_n/n \|f_n\| \to 0$ as $n \to \infty$, which is contrary to the hypothesis that K is continuous.

We see that the linear integral operator (1), with a square integrable kernel, is continuous.

If K and L are two bounded linear operators which map H onto itself, then the product operator LK corresponds to

$$(LK)f = L(Kf).$$

Then

$$\|LKf\| \leqslant \|L\| \|Kf\| \leqslant \|L\| \|K\| \|f\|$$

so that

$$\|LK\| \leqslant \|L\| \|K\| \tag{32}$$

and therefore LK is a bounded operator.

Also

$$(LKf, g) = (Kf, L^*g)$$
$$= (f, K^*L^*g)$$

and hence

$$(LK)^* = K^*L^*. \tag{33}$$

Lastly it is interesting to observe that the norms of bounded linear operators K and L satisfy the triangle inequality. For, using the triangle inequality (6.58) for the elements of Hilbert space, we have

$$\|Kf + Lf\| \leq \|Kf\| + \|Lf\|$$
$$\leq (\|K\| + \|L\|)\|f\|$$

so that

$$\|K + L\| \leq \|K\| + \|L\|. \tag{34}$$

7.2.1 Matrix representation

Suppose that $\phi_1, \phi_2, \ldots, \phi_r, \ldots$ is a complete orthonormal system or basis in H. Then the matrix with elements

$$k_{rs} = (K\phi_r, \phi_s) \tag{35}$$

is called the *kernel matrix* of the bounded operator K.

Introducing the Fourier coefficients

$$x_r = (f, \phi_r), \qquad y_s = (g, \phi_s) \tag{36}$$

where $f, g \in H$, we have

$$f = \sum_{r=1}^{\infty} x_r \phi_r, \qquad g = \sum_{s=1}^{\infty} y_s \phi_s \tag{37}$$

and

$$Kf = \sum_{r=1}^{\infty} x_r K\phi_r = \sum_{r=1}^{\infty} \sum_{s=1}^{\infty} x_r k_{rs} \phi_s$$

so that (Kf, g) has the bilinear form

$$(Kf, g) = \sum_{r=1}^{\infty} \sum_{s=1}^{\infty} x_r k_{rs} \bar{y}_s. \tag{38}$$

Since K is bounded it follows that

$$|(Kf, g)| \leq \|K\| \|f\| \|g\|$$
$$= \|K\| \sqrt{\left(\sum_{r=1}^{\infty} |x_r|^2 \right) \left(\sum_{s=1}^{\infty} |y_s|^2 \right)}.$$

Hence

$$\left| \sum_{r=1}^{\infty} \sum_{s=1}^{\infty} x_r k_{rs} \bar{y}_s \right| \leqslant \|K\| \sqrt{\left(\sum_{r=1}^{\infty} |x_r|^2 \right) \left(\sum_{s=1}^{\infty} |y_s|^2 \right)} \tag{39}$$

and thus the bilinear form (38) is also bounded.

7.3 Completely continuous operators

We now introduce the concept of *weak convergence*. A sequence of elements $\{f_n\}$ in Hilbert space is said to converge weakly to a limit element f if $(f_n, g) \to (f, g)$ as $n \to \infty$ for all $g \in H$. Weak convergence is written $f_n \rightharpoonup f$ as $n \to \infty$.

If a sequence is strongly convergent it is also weakly convergent. For, by Schwarz's inequality (6.57), we have

$$|(f_n - f, g)| \leqslant \|f_n - f\| \|g\|$$

and thus if $\|f_n - f\| \to 0$ as $n \to \infty$ then $(f_n, g) \to (f, g)$ as $n \to \infty$. However, the converse need not be true. Thus we may have $f_n \rightharpoonup f$ but $f_n \nrightarrow f$ as $n \to \infty$.

Let us suppose that K is a bounded linear operator in H. Then

$$|(Kf_n - Kf, g)| \leqslant \|K(f_n - f)\| \|g\|$$
$$\leqslant \|K\| \|f_n - f\| \|g\|$$

and so if $f_n \to f$ as $n \to \infty$ it follows that $Kf_n \rightharpoonup Kf$, that is Kf_n is weakly convergent to Kf.

Now suppose that $Kf_n \to Kf$ when $f_n \rightharpoonup f$ as $n \to \infty$. Then K is called a *completely continuous* (or compact) linear operator in H.

Every completely continuous operator is bounded. For if $f_n \to f$ as $n \to \infty$ we have shown above that $f_n \rightharpoonup f$ and so $Kf_n \to Kf$. Thus a completely continuous operator is a continuous operator and this means that it is bounded as shown in section 7.2.

However a continuous operator is not necessarily completely continuous. For example the identity operator I is continuous but it is not completely continuous for if $f_n \rightharpoonup f$ but $f_n \nrightarrow f$ as $n \to \infty$ then obviously $If_n \rightharpoonup If$ but $If_n \nrightarrow If$.

If the kernel matrix (k_{rs}) of a bounded linear operator K satisfies

$$\sum_{r=1}^{\infty} \sum_{s=1}^{\infty} |k_{rs}|^2 < \infty \tag{40}$$

then K is completely continuous. For we have, using (38), that

$$(K(f_n - f), \phi_s) = \sum_{r=1}^{\infty} (x_r^{(n)} - x_r) k_{rs} \tag{41}$$

where $x_r^{(n)}$ and x_r are the Fourier coefficients of f_n and f respectively, which gives

$$\|K(f_n - f)\|^2 \leq \sum_{s=1}^{m} |(K(f_n - f), \phi_s)|^2$$
$$+ \sum_{s=m+1}^{\infty} \sum_{r=1}^{\infty} |k_{rs}|^2 \|f_n - f\|^2 \tag{42}$$

using Cauchy's inequality.

On the right-hand side of (42) the second sum approaches zero as $m \to \infty$ uniformly in n since $\|f_n - f\|$ is bounded and the operator K satisfies the condition (40). The first sum approaches zero as $n \to \infty$ since $(K(f_n - f), \phi_s) = (f_n - f, K^* \phi_s) \to 0$ if $f_n \rightharpoonup f$. Thus $Kf_n \to Kf$ if $f_n \rightharpoonup f$ as $n \to \infty$ and thus K is completely continuous.

Any finite dimensional linear operator is completely continuous since for this case the double sum on the left-hand side of (40) contains a finite number of terms only.

7.3.1 Integral operator with square integrable kernel

As an example of a completely continuous operator we consider the integral operator K over the space of square integrable functions given by

$$g(x) = \int_a^b K(x, s) f(s) \, ds \qquad (a \leq x \leq b) \tag{43}$$

where the kernel is square integrable and satisfies

$$\left. \begin{array}{ll} \displaystyle\int_a^b |K(x, s)|^2 \, ds < \infty & (a \leq x \leq b), \\[3mm] \displaystyle\int_a^b |K(x, s)|^2 \, dx < \infty & (a \leq s \leq b), \\[3mm] \displaystyle\int_a^b \int_a^b |K(x, s)|^2 \, dx \, ds < \infty. \end{array} \right\} \tag{44}$$

By Schwarz's inequality we have

$$|g(x)|^2 \le \int_a^b |K(x, s)|^2 \, ds \int_a^b |f(s)|^2 \, ds$$

and so

$$\int_a^b |g(x)|^2 \, dx \le \int_a^b \int_a^b |K(x, s)|^2 \, ds \, dx \int_a^b |f(s)|^2 \, ds,$$

that is

$$\|g\|^2 \le \int_a^b \int_a^b |\boldsymbol{K}(x, s)|^2 \, ds \, dx \, \|f\|^2. \tag{45}$$

Hence

$$\|\boldsymbol{K}\| \le \sqrt{\int_a^b \int_a^b |K(x, s)|^2 \, ds \, dx} = \|\boldsymbol{K}\|_2 \tag{46}$$

since the norm of $\boldsymbol{K}$ is the least upper bound of $\|g\|/\|f\|$.

Now consider any complete orthonormal system $\phi_1(x), \phi_2(x), \ldots, \phi_r(x), \ldots$ in the Hilbert space of L^2 functions defined over $a \le x \le b$. We let

$$k_{rs} = \int_a^b \int_a^b \overline{\phi_s(x)} K(x, t) \phi_r(t) \, dx \, dt, \tag{47}$$

$$x_r = \int_a^b f(t) \overline{\phi_r(t)} \, dt \tag{48}$$

and

$$y_s = \int_a^b g(x) \overline{\phi_s(x)} \, dx. \tag{49}$$

Then

$$y_s = \int_a^b \int_a^b K(x, t) f(t) \overline{\phi_s(x)} \, dx \, dt \tag{50}$$

But

$$\int_a^b K(x, t) \overline{\phi_s(x)} \, dx = \sum_{r=1}^\infty k_{rs} \overline{\phi_r(t)} \tag{51}$$

and

$$f(t) = \sum_{r=1}^\infty x_r \phi_r(t) \tag{52}$$

for almost all values of t, that is except for a set of values of t of Lebesgue measure zero, and hence

$$y_s = \sum_{r=1}^{\infty} x_r k_{rs}. \tag{53}$$

Also

$$\int_a^b \left| \int_a^b K(x, t) \overline{\phi_s(x)} \, dx \right|^2 dt = \sum_{r=1}^{\infty} |k_{rs}|^2 \tag{54}$$

and, using Parseval's formula (6.66),

$$\int_a^b |K(x, t)|^2 \, dx = \sum_{s=1}^{\infty} \left| \int_a^b K(x, t) \overline{\phi_s(x)} \, dx \right|^2. \tag{55}$$

Consequently

$$\int_a^b \int_a^b |K(x, t)|^2 \, dx \, dt = \sum_{r=1}^{\infty} \sum_{s=1}^{\infty} |k_{rs}|^2 \tag{56}$$

and so the matrix representation of the integral operator K with square integrable kernel satisfies the condition (40) which ensures that K is completely continuous.

Problems

1. If l_2 is the Hilbert space of infinite sequences

$$\mathbf{x} = (x_1, x_2, \ldots, x_r, \ldots)$$

and K is the operator defined by

$$K\mathbf{x} = (0, x_2, x_3, \ldots, x_r, \ldots),$$

show that K is a linear operator in l_2 having unit norm.

2. If K is the operator in l_2 defined by

$$K\mathbf{x} = \mathbf{y}$$

where $y_r = x_{r+1}/(r+1) \, (r \geq 1)$, show that K is a bounded linear operator and find $\|K\|$.

3. Prove that the linear differential operator $L = d/dx$, acting on the space of continuous functions $f(x)$ defined in the interval $0 \leq x \leq 1$, is unbounded by showing that $\|Lf_n\| = n\pi$ where $f_n(x) = \sqrt{2} \sin n\pi x$.

4. Show that

(i) the integral operator

$$\boldsymbol{K}_1 = \int_0^1 \frac{\sin xs}{s}\, ds \qquad (0 \leqslant x \leqslant 1)$$

is bounded, but that

(ii) the integral operator

$$\boldsymbol{K}_2 = \int_0^1 \frac{\sin xs}{s^2}\, ds \qquad (0 \leqslant x \leqslant 1)$$

is unbounded.

5. A *normal operator K* satisfies

$$\boldsymbol{K}\boldsymbol{K}^* = \boldsymbol{K}^*\boldsymbol{K}.$$

Show that the normal operator $\boldsymbol{K}$ also satisfies

$$\|\boldsymbol{K}f\| = \|\boldsymbol{K}_r^* f\|$$

for every element $f \in H$.

$$\text{If } \boldsymbol{M} = (\boldsymbol{K} + \boldsymbol{K}^*)/2, \qquad \boldsymbol{N} = (\boldsymbol{K} - \boldsymbol{K}^*)/2i$$

show that the operator $\boldsymbol{K}$ is normal if and only if $\boldsymbol{M}$ and $\boldsymbol{N}$ commute, that is $\boldsymbol{M}\boldsymbol{N} = \boldsymbol{N}\boldsymbol{M}$.

6. Find the kernel matrix (k_{rs}) for

$$K(x, t) = \cos(x - t) \qquad (-\pi \leqslant x \leqslant \pi, -\pi \leqslant t \leqslant \pi)$$

using the orthonormal basis composed of the trigonometric functions

$$\left\{ \frac{1}{\sqrt{2\pi}}, \frac{1}{\sqrt{\pi}} \cos rx, \frac{1}{\sqrt{\pi}} \sin rx \right\} \qquad (r = 1, 2, \ldots)$$

defined in the interval $-\pi \leqslant x \leqslant \pi$.

Show that

$$\int_{-\pi}^{\pi} \int_{-\pi}^{\pi} |K(x, t)|^2\, dx\, dt = \sum_r \sum_s |k_{rs}|^2 = 2\pi^2.$$

7. Find the kernel matrix (k_{rs}) for

$$K(x, t) = x^2 + t^2 \qquad (-1 \leqslant x \leqslant 1, -1 \leqslant t \leqslant 1)$$

using the orthonormal basis composed of the functions

$$\phi_r(x) = \sqrt{\frac{2r+1}{2}} \, P_r(x) \qquad (r = 0, 1, 2, \ldots)$$

where the $P_r(x)$ are Legendre polynomials defined in the interval $-1 \leqslant x \leqslant 1$.

Show that

$$\int_{-1}^{1} \int_{-1}^{1} |K(x, t)|^2 \, dx \, dt = \sum_{r=0}^{\infty} \sum_{s=0}^{\infty} |k_{rs}|^2 = \frac{112}{45}.$$

The resolvent

We have already indicated in previous chapters that the solution of an integral equation of the second kind can be expressed in terms of a resolvent kernel. The purpose of the present chapter is to examine the resolvent kernel and the resolvent operator in some detail and to derive results of greater generality than before.

8.1   Resolvent equation

Let us turn our attention again to the Fredholm linear integral equation of the second kind

$$\phi(x) = f(x) + \lambda \int_a^b K(x, s)\phi(s)\,\mathrm{d}s \qquad (a \leqslant x \leqslant b) \qquad (1)$$

where λ is a parameter. Introducing the integral operator

$$\boldsymbol{K} = \int_a^b K(x, s)\,\mathrm{d}s \qquad (2)$$

we may rewrite this equation in the form

$$\phi = f + \lambda \boldsymbol{K}\phi \qquad (3)$$

or

$$(\boldsymbol{I} - \lambda \boldsymbol{K})\phi = f. \qquad (4)$$

We now seek an integral operator $\boldsymbol{R}$ given by

$$\boldsymbol{R} = \int_a^b R(x, s; \lambda)\,\mathrm{d}s \qquad (5)$$

such that

$$\phi = f + \lambda \boldsymbol{R}f, \qquad (6)$$

that is

$$\phi = (\boldsymbol{I} + \lambda \boldsymbol{R})f. \qquad (7)$$

The operator $\boldsymbol{R}$ depends on the parameter λ and since it provides the solution to the integral equation (1), R is called the resolvent and $R(x, s; \lambda)$ the resolvent kernel. We have seen in section 1.3.3 how the resolvent kernel arises in a natural way for integral equations possessing separable kernels.

Substituting (7) into (4) we find that, provided R exists, it must satisfy

$$(I - \lambda K)(I + \lambda R)f = f \qquad (8)$$

and so we anticipate that

$$R - K = \lambda KR. \qquad (9)$$

Likewise substituting (4) into (7) we find that

$$\phi = (I + \lambda R)(I - \lambda K)\phi \qquad (10)$$

which yields

$$R - K = \lambda RK. \qquad (11)$$

The operator equation

$$R - K = \lambda KR = \lambda RK \qquad (12)$$

is called the *resolvent equation*.

If there exists an operator R with a square integrable kernel $R(x, s; \lambda)$ satisfying the resolvent equation (12) for a given value of λ, then this value of λ is said to be a *regular value* of the kernel $K(x, s)$. The set of all regular values of an operator K is known as the *resolvent set* Λ.

The *adjoint equation* of (3) is defined as

$$\psi = g + \bar{\lambda} K^* \psi \qquad (13)$$

where K^* is the adjoint operator of K and g is a given square integrable function while ψ is a square integrable solution.

Taking adjoints of the operators occurring in the resolvent equation (12) we obtain

$$R^* - K^* = \bar{\lambda} R^* K^* = \bar{\lambda} K^* R^* \qquad (14)$$

and so R^* is the resolvent for the adjoint equation (13).

Evidently $\bar{\lambda}$ is a regular value of K^* if and only if λ is a regular value of K.

8.2   Uniqueness theorem

We shall show first in this section that if there exists a resolvent kernel $R(x, s; \lambda)$ of the kernel $K(x, s)$ for a given value of the parameter λ, then it is unique.

For suppose that there are two such kernels $R_1(x, s; \lambda)$ and $R_2(x, s; \lambda)$. Then

$$\boldsymbol{R}_1 - \boldsymbol{K} = \lambda \boldsymbol{R}_1 \boldsymbol{K} = \lambda \boldsymbol{K} \boldsymbol{R}_1$$
$$\boldsymbol{R}_2 - \boldsymbol{K} = \lambda \boldsymbol{R}_2 \boldsymbol{K} = \lambda \boldsymbol{K} \boldsymbol{R}_2$$

and so, putting

$$\boldsymbol{\Gamma} = \boldsymbol{R}_1 - \boldsymbol{R}_2 \tag{15}$$

we see that

$$\boldsymbol{\Gamma} = \lambda \boldsymbol{K} \boldsymbol{\Gamma}. \tag{16}$$

Hence

$$\begin{aligned}
\boldsymbol{R}_1 \boldsymbol{\Gamma} &= \lambda \boldsymbol{R}_1 \boldsymbol{K} \boldsymbol{\Gamma} \\
&= (\boldsymbol{R}_1 - \boldsymbol{K}) \boldsymbol{\Gamma} \\
&= \boldsymbol{R}_1 \boldsymbol{\Gamma} - \boldsymbol{K} \boldsymbol{\Gamma}
\end{aligned}$$

yielding

$$\boldsymbol{K} \boldsymbol{\Gamma} = 0. \tag{17}$$

It follows from (16) that $\boldsymbol{\Gamma} = 0$ giving

$$\boldsymbol{R}_1 = \boldsymbol{R}_2. \tag{18}$$

Next we show that if $f(x)$ is a square integrable function and if λ is a regular value of the square integrable kernel $K(x, s)$ possessing the square integrable resolvent kernel $R(x, s; \lambda)$, then the integral equation (3) has the unique square integrable solution (6).

Suppose that the function $\phi(x)$ is given by (6). Then we have

$$\begin{aligned}
f + \lambda \boldsymbol{K} \phi &= f + \lambda \boldsymbol{K}(f + \lambda \boldsymbol{R} f) \\
&= f + \lambda \boldsymbol{K} f + \lambda^2 \boldsymbol{K} \boldsymbol{R} f \\
&= f + \lambda \boldsymbol{K} f + \lambda (\boldsymbol{R} - \boldsymbol{K}) f \\
&= f + \lambda \boldsymbol{R} f \\
&= \phi
\end{aligned}$$

and so ϕ is a solution of the integral equation (3).

Conversely, if the square integrable function ϕ satisfies (3) we have

$$f = \phi - \lambda \boldsymbol{K} \phi$$

so that

$$f + \lambda \boldsymbol{R}f = \phi - \lambda \boldsymbol{K}\phi + \lambda \boldsymbol{R}(\phi - \lambda \boldsymbol{K}\phi)$$
$$= \phi + \lambda(\boldsymbol{R} - \boldsymbol{K} - \lambda \boldsymbol{RK})\phi$$
$$= \phi$$

using the resolvent equation (12), which proves the uniqueness of the solution.

However it should be noted that there may exist other solutions of (3) which are not square integrable.

8.3 Characteristic values and functions

We next consider the homogeneous linear integral equation of the second kind

$$\phi(x) = \lambda \int_a^b K(x, s)\phi(s) \, \mathrm{d}s \tag{19}$$

which we may rewrite in the form

$$\phi = \lambda \boldsymbol{K}\phi \tag{20}$$

where $\boldsymbol{K}$ is the linear integral operator (2).

This equation clearly has the solution $\phi = 0$ for all values of λ. In addition to the solution $\phi = 0$, the integral equation (20) may have other square integrable solutions $\phi_1(x), \phi_2(x), \ldots, \phi_\nu(x), \ldots$ for certain values $\lambda_1, \lambda_2, \ldots, \lambda_\nu, \ldots$ respectively of the parameter λ. We shall call these values of λ characteristic values. However they are sometimes called eigenvalues although the term eigenvalue is usually reserved for the values of λ^{-1}. The corresponding solutions of (20) are called characteristic functions or eigenfunctions.

A regular value λ of a square integrable kernel cannot be a characteristic value. For then we would have

$$\phi = f + \lambda \boldsymbol{K}\phi$$

with $f = 0$. But for a regular value of λ the integral equation has the unique square integrable solution

$$\phi = f + \lambda \boldsymbol{R}f$$

from which it follows immediately that $\phi = 0$.

The complement of the resolvent set Λ is called the *spectrum* $\sum$ of the operator $\boldsymbol{K}$. Thus $\sum$ contains the set of characteristic values of $\boldsymbol{K}$.

If $\phi(x)$ is a square integrable characteristic function of a continuous kernel $K(x, s)$ we have

$$|\phi(x) - \phi(x')| = |\lambda| \left| \int_a^b \{K(x, s) - K(x', s)\}\phi(s)\, ds \right|$$

$$\leqslant |\lambda| \|\phi\| \left\{ \int_a^b |K(x, s) - K(x', s)|^2\, ds \right\}^{1/2}$$

using Schwarz's inequality. Hence, given $\varepsilon > 0$, there exists $\delta > 0$ such that $|\phi(x) - \phi(x')| < \varepsilon$ if $|x - x'| < \delta$ by virtue of the continuity of $K(x, s)$, and thus $\phi(x)$ is also continuous. However if $\lambda^{-1} = 0$ corresponding to a zero eigenvalue the above argument breaks down and the eigenfunction can be discontinuous (see problem 5, p. 112).

Now suppose that ϕ is a characteristic function of the kernel K for the characteristic value λ, and ψ is a characteristic function of the adjoint kernel K^* for the characteristic value $\bar{\mu}$ where $\mu \neq \lambda$. Then

$$(\phi, \psi) = (\lambda \boldsymbol{K}\phi, \psi) = \lambda(\boldsymbol{K}\phi, \psi)$$

and

$$(\phi, \psi) = (\phi, \bar{\mu}\boldsymbol{K}^*\psi) = \mu(\boldsymbol{K}\phi, \psi)$$

using (7.24), so that

$$(\lambda - \mu)(\boldsymbol{K}\phi, \psi) = 0.$$

Since $\mu \neq \lambda$ it follows that $(\boldsymbol{K}\phi, \psi) = 0$ and hence

$$(\phi, \psi) = 0,$$

that is, ϕ and ψ are orthogonal.

8.4 Neumann series

We have already shown in chapter 4 that the method of successive substitutions leads to the solution of the Fredholm linear integral equation of the second kind (1) in the form of an infinite series, named after Neumann, given by

$$\phi(x) = f(x) + \lambda \int_a^b R(x, s; \lambda)f(s)\, ds \tag{21}$$

where

$$R(x, s; \lambda) = \sum_{n=0}^{\infty} \lambda^n K_{n+1}(x, s) \qquad (22)$$

is the resolvent kernel.

In terms of the integral operator (2) it can be readily seen that the solution ϕ to the integral equation (1) can be written

$$\phi = f + \sum_{n=1}^{\infty} \lambda^n \boldsymbol{K}^n f \qquad (23)$$

while the resolvent $\boldsymbol{R}$ can be expressed in the form

$$\boldsymbol{R} = \sum_{n=0}^{\infty} \lambda^n \boldsymbol{K}^{n+1}. \qquad (24)$$

The Neumann series for the solution to the Fredholm equation (1) and for the resolvent kernel (22) can be shown to be convergent under considerably less stringent conditions than those derived in sections 4.1 and 4.2. Thus let us suppose that $f(x)$ is a square integrable function and that $K(x, s)$ is a square integrable kernel satisfying

$$\int_a^b |K(x, t)|^2 \, \mathrm{d}t < A^2 \qquad (a \leq x \leq b) \qquad (25)$$

where A is a finite positive constant. Since

$$K_{n+1}(x, s) = \int_a^b K_n(x, t) K(t, s) \, \mathrm{d}t \qquad (26)$$

where $K_n(x, s)$ is the iterated kernel defined in section 4.2, we have by Schwarz's inequality

$$|K_{n+1}(x, s)|^2 \leq \int_a^b |K_n(x, t)|^2 \, \mathrm{d}t \int_a^b |K(t, s)|^2 \, \mathrm{d}t$$

so that, on integrating over s, we obtain

$$\int_a^b |K_{n+1}(x, s)|^2 \, \mathrm{d}s \leq \int_a^b |K_n(x, t)|^2 \, \mathrm{d}t \, \|\boldsymbol{K}\|_2^2 \qquad (27)$$

where $\|\boldsymbol{K}\|_2$ is the norm of $K(t, s)$ given by (7.17). Repeated application of (27) gives

$$\int_a^b |K_{n+1}(x, s)|^2 \, \mathrm{d}s \leq A^2 \|\boldsymbol{K}\|_2^{2n} \qquad (28)$$

from which it follows that

$$\left| \int_a^b R(x, s; \lambda) f(s) \, ds \right| \leq \sum_{n=0}^{\infty} |\lambda|^n \left| \int_a^b K_{n+1}(x, s) f(s) \, ds \right|$$

$$\leq \sum_{n=0}^{\infty} |\lambda|^n \|f\| \left\{ \int_a^b |K_{n+1}(x, s)|^2 \, ds \right\}^{1/2}$$

$$\leq A \|f\| \sum_{n=0}^{\infty} |\lambda|^n \|K\|_2^n. \tag{29}$$

Hence the Neumann series for $\phi(x)$ converges absolutely and uniformly for $|\lambda| \|K\|_2 < 1$.

If we assume further that

$$\int_a^b |K(t, s)|^2 \, dt < B^2 \qquad (a \leq s \leq b) \tag{30}$$

where B is a finite positive constant, then

$$|R(x, s; \lambda) - K(x, s)| \leq \sum_{n=1}^{\infty} |\lambda|^n |K_{n+1}(x, s)|$$

$$\leq \sum_{n=1}^{\infty} |\lambda|^n \left\{ \int_a^b |K(x, t)|^2 \, dt \right\}^{1/2} \|K\|_2^{n-1} \left\{ \int_a^b |K(t, s)|^2 \, dt \right\}^{1/2}$$

$$\leq AB \sum_{n=1}^{\infty} |\lambda|^n \|K\|_2^{n-1} \tag{31}$$

and so the Neumann series for the resolvent kernel converges absolutely and uniformly if $|\lambda| \|K\|_2 < 1$.

We have shown in section 8.2 that the solution ϕ and the resolvent R are unique. Hence if $|\lambda| \|K\|_2 < 1$ the corresponding homogeneous equation (19) has the unique solution $\phi = 0$. However this can be verified directly for, by Schwarz's inequality, we have

$$|\phi(x)|^2 \leq |\lambda|^2 \|\phi\|^2 \int_a^b |K(x, s)|^2 \, ds$$

giving, on performing an integration over x,

$$\|\phi\|^2 \leq |\lambda|^2 \|\phi\|^2 \|K\|_2^2$$

i.e.

$$(1 - |\lambda|^2 \|K\|_2^2) \|\phi\|^2 \leq 0.$$

But $|\lambda| \|K\|_2 < 1$ and so it follows that $\|\phi\| = 0$ which implies $\phi = 0$.

This also demonstrates that the solution of the integral equation (1) is unique when $|\lambda| \|K\|_2 < 1$. For if there exist two solutions $\phi_1(x)$, $\phi_2(x)$ satisfying

$$\phi_1(x) = f(x) + \lambda \int_a^b K(x, s)\phi_1(s)\,\mathrm{d}s$$

$$\phi_2(x) = f(x) + \lambda \int_a^b K(x, s)\phi_2(s)\,\mathrm{d}s,$$

we see at once that $\phi(x) = \phi_1(x) - \phi_2(x)$ is a solution of the homogeneous equation (19) from which it follows from the preceding result that $\phi(x) = 0$, giving $\phi_1(x) = \phi_2(x)$.

8.4.1 Volterra integral equation of the second kind

We now examine the convergence of the Neumann series solution of the Volterra linear integral equation of the second kind

$$\phi(x) = f(x) + \lambda \int_a^x K(x, s)\phi(s)\,\mathrm{d}s \qquad (a \leqslant x \leqslant b) \qquad (32)$$

where $f(x)$ is a square integrable function and $K(x, s)$ is a square integrable kernel satisfying $K(x, s) = 0$ for $a \leqslant x < s \leqslant b$, and

$$\alpha^2(x) = \int_a^x |K(x, s)|^2\,\mathrm{d}s < A^2 \qquad (a \leqslant x \leqslant b) \qquad (33)$$

$$\beta^2(s) = \int_s^b |K(x, s)|^2\,\mathrm{d}x < B^2 \qquad (a \leqslant s \leqslant b) \qquad (34)$$

where A and B are positive constants.

We shall show that the Neumann series for the resolvent kernel is absolutely and uniformly convergent for all λ.

We begin by noting that, from Schwarz's inequality,

$$\begin{aligned}
|K_2(x, s)|^2 &= \left| \int_s^x K(x, t)K(t, s)\,\mathrm{d}t \right|^2 \\
&\leqslant \int_s^x |K(x, t)|^2\,\mathrm{d}t \int_s^x |K(t, s)|^2\,\mathrm{d}t \\
&\leqslant \alpha^2(x)\beta^2(s)
\end{aligned} \qquad (35)$$

and

$$|K_3(x, s)|^2 = \left| \int_s^x K(x, t) K_2(t, s) \, dt \right|^2$$

$$\leq \int_s^x |K(x, t)|^2 \, dt \int_s^x |K_2(t, s)|^2 \, dt$$

$$\leq \alpha^2(x) \beta^2(s) \int_s^x \alpha^2(t) \, dt,$$

i.e.,

$$|K_3(x, s)|^2 \leq \alpha^2(x) \beta^2(s) k_1(x, s) \tag{36}$$

where

$$k_1(x, s) = \int_s^x \alpha^2(t) \, dt. \tag{37}$$

Next we use the principle of induction to establish the inequality

$$|K_{n+2}(x, s)|^2 \leq \alpha^2(x) \beta^2(s) k_n(x, s) \qquad (n = 1, 2, \ldots) \tag{38}$$

where

$$k_n(x, s) = \int_s^x \alpha^2(t) k_{n-1}(t, s) \, dt \qquad (n = 2, 3, \ldots) \tag{39}$$

We have already proved that the inequality (38) holds for $n = 1$. Now assuming the truth of the inequality for a positive integer n and using Schwarz's inequality again, we see that

$$|K_{n+3}(x, s)|^2 = \left| \int_s^x K(x, t) K_{n+2}(t, s) \, dt \right|^2$$

$$\leq \int_s^x |K(x, t)|^2 \, dt \int_s^x |K_{n+2}(t, s)|^2 \, dt$$

$$\leq \alpha^2(x) \beta^2(s) \int_s^x \alpha^2(t) k_n(t, s) \, dt$$

$$= \alpha^2(x) \beta^2(s) k_{n+1}(x, s)$$

which establishes the inequality for integer $n + 1$.

Also, by using the principle of induction, we can show that

$$k_n(x, s) = \frac{\{k_1(x, s)\}^n}{n!}. \tag{40}$$

Obviously this holds for $n = 1$. Assuming its validity for a positive integer n we have, using the definition (39), that

$$
\begin{aligned}
k_{n+1}(x, s) &= \frac{1}{n!} \int_s^x \alpha^2(t) \{k_1(t, s)\}^n \, dt \\
&= \frac{1}{n!} \int_s^x \left\{ \frac{\partial}{\partial t} k_1(t, s) \right\} \{k_1(t, s)\}^n \, dt \\
&= \frac{1}{n!} \left[\frac{\{k_1(t, s)\}^{n+1}}{n+1} \right]_s^x \\
&= \frac{\{k_1(x, s)\}^{n+1}}{(n+1)!}
\end{aligned}
$$

and so the formula (40) holds for integer $n + 1$.

Now

$$
k_1(x, s) = \int_s^x dt \int_a^t |K(t, s)|^2 \, ds \le \|K\|_2^2 \tag{41}
$$

and so, using (38), (33), (34), (40) and (41) we obtain

$$
|K_{n+2}(x, s)| \le \frac{AB \|K\|_2^n}{\sqrt{n!}}. \tag{42}
$$

Hence

$$
\begin{aligned}
|R(x, s; \lambda) - K(x, s)| &\le \sum_{n=0}^\infty |\lambda|^{n+1} |K_{n+2}(x, s)| \\
&\le \sum_{n=0}^\infty \frac{|\lambda|^{n+1} AB \|K\|_2^n}{\sqrt{n!}} \\
&= |\lambda| \, AB \sum_{n=0}^\infty \frac{|\lambda|^n \|K\|_2^n}{\sqrt{n!}} \tag{43}
\end{aligned}
$$

which shows that the Neumann series for the resolvent kernel is absolutely and uniformly convergent for all λ since the infinite series on the right-hand side of (43) is of the type $\sum_{n=0}^\infty z^n / \sqrt{n!}$ which converges for $|z| < \infty$ by the ratio test. Thus every value of λ is a regular value for the Volterra equation of the second kind.

Since a regular value of λ cannot be a characteristic value, it follows that the homogeneous equation

$$\phi(x) = \lambda \int_a^x K(x, s)\phi(s)\, ds \qquad (a \leq x \leq b) \tag{44}$$

has the unique square integrable solution $\phi = 0$ for all λ.

The Neumann series solution of the Volterra equation of the second kind (32) takes the form

$$\phi(x) = f(x) + \lambda \sum_{n=0}^{\infty} \lambda^n f_{n+1}(x) \tag{45}$$

where

$$f_n(x) = \int_a^x K_n(x, s) f(s)\, ds. \tag{46}$$

We have, by Schwarz's inequality, that

$$\begin{aligned}
|f_1(x)|^2 &\leq \int_a^x |K(x, s)|^2\, ds \int_a^x |f(s)|^2\, ds \\
&\leq A^2 \|f\|^2
\end{aligned} \tag{47}$$

and, using the inequalities (38) and (41) together with (33) and (40), that

$$\begin{aligned}
|f_{n+1}(x)|^2 &\leq \int_a^x |K_{n+1}(x, s)|^2\, ds \int_a^s |f(s)|^2\, ds \\
&\leq \frac{A^2 \|\boldsymbol{K}\|_2^{2(n-1)}}{(n-1)!} \int_a^x \beta^2(s)\, ds \, \|f\|^2 \\
&\leq \frac{A^2 \|f\|^2 \|\boldsymbol{K}\|_2^{2n}}{(n-1)!}.
\end{aligned} \tag{48}$$

Hence

$$\begin{aligned}
|\phi(x) - f(x)| &\leq |\lambda| \sum_{n=0}^{\infty} |\lambda|^n |f_{n+1}(x)| \\
&\leq |\lambda| \, A \, \|f\| \left\{ 1 + \sum_{n=1}^{\infty} \frac{|\lambda|^n \|\boldsymbol{K}\|_2^n}{\sqrt{(n-1)!}} \right\}
\end{aligned} \tag{49}$$

and thus the Neumann series for $\phi(x)$ is absolutely and uniformly convergent for $|\lambda| < \infty$.

This demonstrates that the Volterra equation of the second kind (32) has a unique square integrable solution $\phi(x)$ for any square integrable function $f(x)$ and all values of λ.

8.4.2 Bôcher's example

Although we have shown that the homogeneous Volterra equation (44) with a square integrable kernel has the unique square integrable solution $\phi = 0$ for all values of the parameter λ it may have other solutions which are not square integrable. Thus consider the homogeneous Volterra equation

$$\phi(x) = \int_0^x s^{x-s} \phi(s) \, ds \qquad (0 \leq x \leq 1) \tag{50}$$

with the bounded kernel

$$K(x, s) = \begin{cases} s^{x-s} & (0 < s \leq x \leq 1) \\ 0 & (s = 0) \end{cases} \tag{51}$$

given by Bôcher in 1909.

It can be readily verified that it has the discontinuous solution

$$\phi_0(x) = \begin{cases} cx^{x-1} & (0 < x \leq 1) \\ 0 & (x = 0) \end{cases} \tag{52}$$

where c is an arbitrary constant. This solution is not square integrable since

$$\int_0^1 \phi_0^2(x) \, dx$$

evidently diverges owing to the presence of the x^{-2} singularity near the origin.

We now see that if $\phi(x)$ is a solution of

$$\phi(x) = f(x) + \int_0^x s^{x-s} \phi(s) \, ds \qquad (0 \leq x \leq 1), \tag{53}$$

then $\phi(x) + \phi_0(x)$ is also a solution of (53). Thus the solution of (53) is unique if and only if we restrict the solution to be square integrable.

8.5 Fredholm equation in abstract Hilbert space

In the previous sections of this chapter we discussed Fredholm's integral equation (1) in terms of the integral operator K given by

(2). However we may consider a more general form of Fredholm equation

$$\phi = f + \lambda K \phi \tag{54}$$

where K is a completely continuous linear operator which maps abstract Hilbert space H onto itself, and the given element f and the solution ϕ belong to H.

If there exists a continuous linear operator R called the resolvent satisfying the resolvent equation

$$R - K = \lambda K R = \lambda R K \tag{55}$$

for a given value of λ, this value is called a regular value. For such a value of λ the resolvent operator is unique as can be verified by using the method described in section 8.2. Moreover (54) has the unique solution

$$\phi = f + \lambda R f \tag{56}$$

as can be seen by direct substitution.

The adjoint equation of (54) is

$$\psi = g + \bar{\lambda} K^* \psi \tag{57}$$

where K^* is the adjoint of K, and g and ψ belong to H. The resolvent R^* of (57) satisfies the resolvent equation

$$R^* - K^* = \bar{\lambda} R^* K^* = \bar{\lambda} K^* R^*. \tag{58}$$

The homogeneous equation

$$\phi = \lambda K \phi \tag{59}$$

possesses the trivial solution $\phi = 0$ for all λ. It may also have other solutions belonging to H called characteristic vectors $\phi_1, \phi_2, \ldots, \phi_\nu, \ldots$ for certain values $\lambda_1, \lambda_2, \ldots, \lambda_\nu, \ldots$ of λ called characteristic values.

Following the analysis given in section 8.3 we can show that a regular value of λ cannot be a characteristic value. Also

$$(\phi, \psi) = 0 \tag{60}$$

where ϕ is a characteristic vector of K for the characteristic value λ and ψ is a characteristic vector of K^* for the characteristic value $\bar{\mu}$ where $\mu \neq \lambda$.

We see from (55) that the resolvent is given by

$$R = K(I - \lambda K)^{-1} \tag{61}$$

where the inverse L^{-1} of an operator L satisfies

$$L^{-1}L = LL^{-1} = I, \tag{62}$$

I being the identity operator.

The resolvent operator may be expressed in the form of a Neumann expansion

$$R = \sum_{n=0}^{\infty} \lambda^n K^{n+1} \tag{63}$$

where, using the triangle inequality (7.34) for bounded linear operators,

$$\|R\| \leq \sum_{n=0}^{\infty} |\lambda|^n \|K^{n+1}\|$$

$$\leq \sum_{n=0}^{\infty} |\lambda|^n \|K\|^{n+1}$$

since $\|K^{n+1}\| \leq \|K\|^{n+1}$ by successive application of (7.32). Hence R is bounded if $|\lambda| \|K\| < 1$.

We also see that R is completely continuous if $|\lambda| \|K\| < 1$, for we have

$$\|Rf_m - Rf\| = \left\| \sum_{n=0}^{\infty} \lambda^n K^{n+1}(f_m - f) \right\|$$

$$\leq \left\| \sum_{n=0}^{\infty} \lambda^n K^n \right\| \|K(f_m - f)\|$$

$$\to 0$$

as $f_m \to f$ since K is completely continuous and $\sum_{n=0}^{\infty} \lambda^n K^n$ is bounded.

Problems

1. If R_λ, R_μ are the resolvents for a given kernel K corresponding to parameter values λ, μ respectively, show that

$$R_\lambda - R_\mu = (\lambda - \mu) R_\lambda R_\mu$$

2. If the linearly independent functions $\phi_1, \phi_2, \ldots, \phi_n$; $\psi_1, \psi_2, \ldots, \psi_n$ form a *biorthogonal* series so that

$$(\phi_i, \psi_j) = \delta_{ij} (i, j = 1, 2, \ldots, n),$$

show that the kernel

$$K(x, s) = \sum_{i=1}^{n} u_i(x) \overline{v_i(s)} \qquad (a \leq x \leq b, \, a \leq s \leq b) \tag{1}$$

has the characteristic functions $\phi_i(x)$ with characteristic values a_i^{-1} $(i = 1, \ldots, n)$.

Further, use the Neumann expansion to show that the resolvent kernel of $K(x, s)$ converges to

$$R(x, s; \lambda) = \sum_{i=1}^{n} \frac{a_i \phi_i(x) \overline{\psi_i(s)}}{1 - a_i \lambda}$$

if $|a_i \lambda| < 1$ for all i.

Verify that the relation obtained in problem 1 is satisfied.

3. Show that the kernel

$$K(x, s) = \sum_{\nu=0}^{\infty} a_\nu \cos \nu x \cos \nu s \qquad (0 \leqslant x \leqslant \pi, 0 \leqslant s \leqslant \pi),$$

where $\sum_{\nu=0}^{\infty} |a_\nu| < \infty$, has the characteristic functions $1/\sqrt{\pi}$, $\sqrt{2/\pi} \cos \nu x$ $(\nu = 1, 2, \ldots)$ with characteristic values $1/\pi a_0, 2/\pi a_\nu$ respectively.

Use the Neumann expansion to show that for sufficiently small λ the resolvent kernel for $K(x, s)$ is given by

$$R(x, s; \lambda) = \frac{a_0}{1 - \pi a_0 \lambda} + \sum_{\nu=1}^{\infty} \frac{a_\nu \cos \nu x \cos \nu s}{1 - \pi/2 \, a_\nu \lambda}.$$

4. Show that the Poisson kernel

$$K(x, s) = \frac{1}{2\pi} \frac{1 - a^2}{1 + a^2 - 2a \cos(x - s)} \quad (0 \leqslant x \leqslant 2\pi, 0 \leqslant s \leqslant 2\pi),$$

where $|a| < 1$, can be expressed as the Fourier series

$$\frac{1}{2\pi} + \frac{1}{\pi} \sum_{\nu=1}^{\infty} a^\nu \cos \nu(x - s).$$

Obtain the Neumann expansion for the resolvent kernel of $K(x, s)$ and show that it converges to

$$R(x, s; \lambda) = \frac{1}{2\pi} \frac{1}{1 - \lambda} + \frac{1}{\pi} \sum_{\nu=1}^{\infty} \frac{a^\nu}{1 - a^\nu \lambda} \cos \nu(x - s)$$

if $|\lambda| < 1$.

5. Show that the continuous kernel

$$K(x, s) = x\{n - 1 - (2n - 1)s\} \qquad (0 \leqslant x \leqslant 1, 0 \leqslant s \leqslant 1)$$

has the discontinuous square integrable eigenfunction when $n > 2$:

$$\phi(x) = \begin{cases} x^{-1/n} & (0 < x \leq 1) \\ 0 & (x = 0) \end{cases}$$

for the eigenvalue 0, that is

$$\int_0^1 K(x, s)\phi(s)\, ds = 0.$$

6. Show that the discontinuous kernel

$$K(x, s) = \begin{cases} 0 & (0 \leq x < \frac{1}{2}, 0 \leq s \leq 1) \\ 1 & (\frac{1}{2} \leq x \leq 1, 0 \leq s \leq 1) \end{cases}$$

has the discontinuous characteristic function

$$\phi(x) = \begin{cases} 0 & (0 \leq x < \frac{1}{2}) \\ 1 & (\frac{1}{2} \leq x \leq 1) \end{cases}$$

for the characteristic value 2.

Use the Neumann expansion to show that the resolvent kernel of $K(x, s)$ is given by

$$R(x, s; \lambda) = \begin{cases} 0 & (0 \leq x < \frac{1}{2}) \\ \dfrac{2}{2 - \lambda} & (\frac{1}{2} \leq x \leq 1) \end{cases}$$

for $|\lambda| < 2$.

7. Show that the discontinuous kernel

$$K(x, s) = \begin{cases} \left(\dfrac{s}{x}\right)^{1/2} & (0 < x \leq 1, 0 \leq s \leq 1) \\ 0 & (x = 0, 0 \leq s \leq 1) \end{cases}$$

and all its iterates $K_n(x, s)$ are not square integrable.

Establish that its resolvent kernel is given by

$$R(x, s; \lambda) = \frac{K(x, s)}{1 - \lambda}.$$

Further show that $K(x, s)$ has the discontinuous characteristic function

$$\phi(x) = \begin{cases} x^{-1/2} & (0 < x \leq 1) \\ 0 & (x = 0) \end{cases}$$

for the characteristic value 1.

9

Fredholm theory

We are now in a suitable position to discuss the theory originally developed by Fredholm in 1903 and by Schmidt in 1907 concerning the solution of linear integral equations. We shall approach this by first considering degenerate kernels for which it is possible to express the resolvent kernel in a closed analytical form. This will enable us to use the method introduced by Schmidt to treat a general square integrable kernel and to establish the Fredholm theorems. We conclude this chapter by obtaining the Fredholm solution for the case of a continuous kernel. This solution expresses the resolvent kernel as the ratio of two power series which are convergent for all values of the parameter λ.

9.1 Degenerate kernels

Let us examine the special case of a degenerate kernel having the separable form

$$K(x, s) = \sum_{i=1}^{n} u_i(x)\overline{v_i(s)} \quad (a \le x \le b, a \le s \le b) \tag{1}$$

where $u_1(x), \ldots, u_n(x)$ and $v_1(s), \ldots, v_n(s)$ are two sets of linearly independent square integrable functions. The least integer n for which a degenerate kernel can be expressed in the form (1) is called the *rank* of the kernel. Thus a degenerate kernel is often called a kernel of *finite rank*. Sometimes such a kernel (1) is referred to as a Pincherle-Goursat kernel.

The corresponding Fredholm linear integral equation of the second kind takes the form

$$\phi(x) = f(x) + \lambda \sum_{i=1}^{n} u_i(x) \int_a^b \overline{v_i(s)}\phi(s) \, ds. \tag{2}$$

To solve this equation we introduce the unknown constants

$$c_i = \int_a^b \overline{v_i(s)}\phi(s)\,\mathrm{d}s \qquad (3)$$

depending on the solution $\phi(x)$, so that we have

$$\phi(x) = f(x) + \lambda \sum_{i=1}^n c_i u_i(x). \qquad (4)$$

On substituting back into (2) we obtain

$$\sum_{i=1}^n u_i(x)\left[c_i - \int_a^b \overline{v_i(s)}\left\{ f(s) + \lambda \sum_{j=1}^n c_j u_j(s) \right\}\mathrm{d}s \right] = 0$$

and since the $u_i(x)$ are linearly independent functions it follows that

$$c_i - \int_a^b \overline{v_i(s)}\left\{ f(s) + \lambda \sum_{j=1}^n c_j u_j(s) \right\}\mathrm{d}s = 0. \qquad (5)$$

Writing

$$a_{ij} = \int_a^b \overline{v_i(s)}u_j(s)\,\mathrm{d}s \qquad (6)$$

and

$$f_i = \int_a^b \overline{v_i(s)}f(s)\,\mathrm{d}s \qquad (7)$$

we obtain the system of linear algebraic equations

$$c_i - \lambda \sum_{j=1}^n a_{ij}c_j = f_i \qquad (i = 1, \ldots, n) \qquad (8)$$

characterized by the determinant

$$d(\lambda) = \det(\mathbf{I} - \lambda\mathbf{A})$$

$$= \begin{vmatrix} 1 - \lambda a_{11} & -\lambda a_{12} & \ldots & -\lambda a_{1n} \\ -\lambda a_{21} & 1 - \lambda a_{22} & \ldots & -\lambda a_{2n} \\ \cdot & \cdot & & \cdot \\ \cdot & \cdot & & \cdot \\ \cdot & \cdot & & \cdot \\ -\lambda a_{n1} & -\lambda a_{n2} & \ldots & 1 - \lambda a_{nn} \end{vmatrix} \qquad (9)$$

where $\mathbf{A}$ is the matrix with elements (a_{ij}) and $\mathbf{I}$ is the unit matrix.

The determinant $d(\lambda)$ is a polynomial of degree n in λ. If $d(\lambda) \neq 0$ for a given value of λ, the system of equations (8) has the unique solution given by Cramer's rule

$$c_i = \frac{1}{d(\lambda)} \sum_{j=1}^{n} d_{ij}(\lambda) f_j \qquad (i = 1, \ldots, n) \tag{10}$$

where (d_{ij}) are the elements of the *adjugate matrix* $\mathbf{D}$ of $\mathbf{I} - \lambda \mathbf{A}$, that is the transposed matrix of its cofactors.

Using (4) we see that

$$\phi(x) = f(x) + \frac{\lambda}{d(\lambda)} \sum_{i=1}^{n} \sum_{j=1}^{n} u_i(x) d_{ij}(\lambda) f_j \tag{11}$$

from which it follows that the resolvent kernel is given by

$$R(x, s_1 \lambda) = \frac{1}{d(\lambda)} \sum_{i=1}^{n} \sum_{j=1}^{n} u_i(x) d_{ij}(\lambda) \overline{v_j(s)}. \tag{12}$$

Further, the corresponding homogeneous equation

$$\phi(x) = \lambda \int_a^b K(x, s) \phi(s) \, ds \tag{13}$$

evidently has the unique solution $\phi(x) = 0$ if $d(\lambda) \neq 0$, that is if λ is a regular value of $K(x, s)$.

Now suppose that $d(\lambda) = 0$. Then λ must coincide with one of the characteristic values λ_ν of the homogeneous equation (13). If r is the rank of the matrix $\mathbf{I} - \lambda \mathbf{A}$ and $p = n - r$ so that $1 \leq p \leq n - 1$, the system of n homogeneous equations

$$c_i - \lambda \sum_{j=1}^{n} a_{ij} c_j = 0 \qquad (i = 1, \ldots, n) \tag{14}$$

has p linearly independent solutions $\{c_i^{(k)}\}(k = 1, \ldots, p)$. Thus the homogeneous equation (13) has the general solution

$$\phi(x) = \sum_{k=1}^{p} \alpha_k \phi^{(k)}(x) \tag{15}$$

where p is called the *rank* (or sometimes the *index* to avoid confusion with the previous terminology) of the characteristic value λ, and

$$\phi^{(k)}(x) = \lambda \sum_{i=1}^{n} c_i^{(k)} u_i(x). \tag{16}$$

Since $\mathbf{I} - \bar{\lambda}\mathbf{A}^* = (\mathbf{I} - \lambda\mathbf{A})^*$ it follows that the characteristic value $\bar{\lambda}$ of the adjoint homogeneous equation

$$\psi(x) = \bar{\lambda} \int_a^b \overline{K(s, x)} \psi(s) \, ds \qquad (17)$$

possesses the same rank p, and therefore has the same number of linearly independent solutions, as the homogeneous equation (13).

We see also that if λ is a characteristic value for which the inhomogeneous equation (2) has a particular square integrable solution $\phi_0(x)$, then its general square integrable solution is

$$\phi(x) = \phi_0(x) + \sum_{k=1}^p \alpha_k \phi^{(k)}(x). \qquad (18)$$

Lastly we shall show that there exists a square integrable solution $\phi(x)$ of the inhomogeneous equation (2) for a given value of λ if and only if $f(x)$ is orthogonal to every square integrable solution of the adjoint homogeneous equation (17).

If λ is a regular value of $K(x, s)$ then $\psi = 0$ and the result is obvious.

However if λ is a characteristic value of $K(x, s)$, and writing the inhomogeneous equation (2) in the operator form

$$\phi = f + \lambda \mathbf{K}\phi, \qquad (19)$$

we see that

$$\begin{aligned}(f, \psi) &= (\phi, \psi) - \lambda(\mathbf{K}\phi, \psi) \\ &= (\phi, \psi - \bar{\lambda}\mathbf{K}^*\psi) = 0\end{aligned} \qquad (20)$$

since the adjoint homogeneous equation (17) may be written

$$\psi = \bar{\lambda}\mathbf{K}^*\psi \qquad (21)$$

It follows that a necessary condition for ϕ to be a solution of (19) is that f should be orthogonal to every square integrable solution of (21).

This is also a sufficient condition. Thus if $(f, \psi) = 0$ for all the p linearly independent solutions of the adjoint homogeneous equation, then the n equations (8) can be reduced to $r = n - p$ independent equations possessing r independent solutions where $1 \le r \le n - 1$.

We now discuss two examples of degenerate kernels.

Example 1.   Consider the degenerate kernel of rank 2

$$K(x, s) = \sin (x + s) = \sin x \cos s + \cos x \sin s \qquad (22)$$

with $0 \le x \le 2\pi$, $0 \le s \le 2\pi$.

We have

$$
\begin{aligned}
u_1(x) &= \sin x, & u_2(x) &= \cos x, \\
v_1(s) &= \cos s, & v_2(s) &= \sin s,
\end{aligned} \qquad (23)
$$

and thus

$$a_{11} = a_{22} = \int_0^{2\pi} \sin s \cos s \, ds = 0,$$

$$a_{12} = \int_0^{2\pi} \cos^2 s \, ds = \pi, \qquad (24)$$

$$a_{21} = \int_0^{2\pi} \sin^2 s \, ds = \pi.$$

Hence

$$d(\lambda) = \begin{vmatrix} 1 & -\pi\lambda \\ -\pi\lambda & 1 \end{vmatrix} = 1 - \pi^2\lambda^2 \qquad (25)$$

yielding the characteristic values $\lambda_1 = 1/\pi$, $\lambda_2 = -1/\pi$ and the corresponding orthonormalized characteristic functions

$$\phi_1(x) = \frac{1}{\sqrt{2\pi}} (\sin x + \cos x), \quad \phi_2(x) = \frac{1}{\sqrt{2\pi}} (\sin x - \cos x) \quad (26)$$

of the homogeneous Fredholm equation of the second kind. Symmetric real kernels such as (22), satisfying $K(s, x) = K(x, s)$ always have real characteristic values as we shall show in section 10.2.

The elements of the adjugate matrix **D** are

$$d_{11} = d_{22} = 1, \quad d_{12} = d_{21} = \pi\lambda \qquad (27)$$

giving for the resolvent kernel

$$R(x, s; \lambda) = \frac{\sin (x + s) + \pi\lambda \cos (x - s)}{1 - \pi^2\lambda^2} \qquad (28)$$

and for the unique solution to the Fredholm equation of the second kind (2) when $\lambda \neq \pm 1/\pi$:

$$\phi(x) = f(x) + \frac{\lambda}{1 - \pi^2 \lambda^2} \{(f_1 + \pi \lambda f_2) \sin x + (\pi \lambda f_1 + f_2) \cos x\} \quad (29)$$

where

$$f_1 = \int_0^{2\pi} f(s) \cos s \, ds, \quad f_2 = \int_0^{2\pi} f(s) \sin s \, ds. \quad (30)$$

If $f(s) = 1$ we have $f_1 = f_2 = 0$ and thus $(f, \phi_1) = (f, \phi_2) = 0$ so that $f(x)$ is orthogonal to the solutions $\phi_1(x)$, $\phi_2(x)$ of the adjoint homogeneous equation (17) for the characteristic values $\lambda_1 = 1/\pi$, $\lambda_2 = -1/\pi$. Hence the inhomogeneous equation (2) has the general solutions $\phi(x) = 1 + \alpha \phi_1(x)$, $1 + \alpha \phi_2(x)$ corresponding to the characteristic values λ_1, λ_2 respectively.

If $f(s) = s$ we have $f_1 = 0$, $f_2 = -2\pi$ and thus $(f, \phi_1) = (f, \phi_2) = -\sqrt{2}\pi$. Since the orthogonality condition is not satisfied in this case the inhomogeneous equation (2) has no solutions for $\lambda = \pm 1/\pi$.

Example 2. Next we consider the degenerate kernel of rank 2

$$K(x, s) = x - s \quad (0 \le x \le 1, 0 \le s \le 1) \quad (31)$$

for which

$$u_1(x) = x, u_2(x) = -1, v_1(s) = 1, v_2(s) = s \quad (32)$$

so that

$$a_{11} = -a_{22} = \int_0^1 s \, ds = \frac{1}{2}$$

$$a_{12} = -\int_0^1 ds = -1, a_{21} = \int_0^1 s^2 \, ds = \frac{1}{3}. \quad (33)$$

Then

$$d(\lambda) = \begin{vmatrix} 1 - \frac{1}{2}\lambda & \lambda \\ -\frac{1}{3}\lambda & 1 + \frac{1}{2}\lambda \end{vmatrix} = 1 + \frac{1}{12}\lambda^2 \quad (34)$$

and so the characteristic values given by $d(\lambda) = 0$ are the pure imaginary numbers $\pm i2\sqrt{3}$. Skew symmetric real kernels such as

(31), satisfying $K(s, x) = -K(x, s)$, always have pure imaginary characteristic values.

The elements of the adjugate matrix **D** of $I - \lambda A$ are

$$d_{11} = 1 + \tfrac{1}{2}\lambda, \; d_{22} = 1 - \tfrac{1}{2}\lambda, \; d_{12} = -\lambda, \; d_{21} = \tfrac{1}{3}\lambda \tag{35}$$

and so the resolvent kernel is

$$R(x, s; \lambda) = \frac{(1 + \tfrac{1}{2}\lambda)x - (1 - \tfrac{1}{2}\lambda)s - \lambda xs - \tfrac{1}{3}\lambda}{1 + \tfrac{1}{12}\lambda^2} \tag{36}$$

giving for the unique solution of (2)

$$\phi(x) = f(x) + \frac{\lambda}{1 + \tfrac{1}{12}\lambda^2}\{(1 + \tfrac{1}{2}\lambda)xf_1 - (1 - \tfrac{1}{2}\lambda)f_2 - \lambda xf_2 - \tfrac{1}{3}\lambda f_1\} \tag{37}$$

valid for all values of $\lambda \neq \pm i2\sqrt{3}$ and in particular for all real λ.

9.2   Approximation by degenerate kernels

In this section our aim is to show that we can approximate a square integrable kernel as closely as we please, in the mean square sense, by a degenerate kernel of sufficiently high rank.

We suppose that $K(x, s)$ is a square integrable kernel satisfying

$$\left.\begin{array}{l} \displaystyle\int_a^b |K(x, s)|^2 \, \mathrm{d}s < \infty \qquad (a \leqslant x \leqslant b), \\[3mm] \displaystyle\int_a^b |K(x, s)|^2 \, \mathrm{d}x < \infty \qquad (a \leqslant s \leqslant b), \\[3mm] \|K\|_2^2 = \displaystyle\int_a^b \int_a^b |K(x, s)|^2 \, \mathrm{d}x \, \mathrm{d}s < \infty, \end{array}\right\} \tag{38}$$

and we shall prove that given any $\varepsilon > 0$ there exists a degenerate kernel $P(x, s)$ such that

$$\|K - P\|_2 < \varepsilon. \tag{39}$$

Let $v_1(s), v_2(s), \ldots, v_i(s), \ldots$ be a complete orthonormal system of square integrable functions and let

$$\int_a^b K(x, s)v_i(s) \, \mathrm{d}s = u_i(x). \tag{40}$$

Using Parseval's formula we see that

$$\sum_{i=1}^{\infty} |u_i(x)|^2 = \int_a^b |K(x, s)|^2 \, ds \tag{41}$$

and so

$$\sum_{i=1}^{\infty} \int_a^b |u_i(x)|^2 \, dx = \|K\|_2^2 \tag{42}$$

from which it follows that there exists a N such that for $n > N$

$$\sum_{i=n+1}^{\infty} \int_a^b |u_i(x)|^2 \, dx < \varepsilon^2. \tag{43}$$

Now, as a consequence of the completeness of the orthonormal system $\{v_i(s)\}$, we have

$$\int_a^b |K(x, s) - \sum_{i=1}^{n} u_i(x)\overline{v_i(s)}|^2 \, ds = \sum_{i=n+1}^{\infty} |u_i(x)|^2$$

and so

$$\int_a^b \int_a^b \left| K(x, s) - \sum_{i=1}^{n} u_i(x)\overline{v_i(s)} \right|^2 \, dx \, ds < \varepsilon^2, \tag{44}$$

Hence taking

$$P(x, s) = \sum_{i=1}^{n} u_i(x)\overline{v_i(s)} \tag{45}$$

where $n > N$, we obtain the desired result (39).

9.3   Fredholm theorems

We consider the Fredholm integral equation of the second kind

$$\phi(x) = f(x) + \lambda \int_a^b K(x, s)\phi(s) \, ds \tag{46}$$

where $K(x, s)$ is a square integrable kernel and λ is chosen so that $|\lambda| < \omega$ where ω is a positive number. From the discussion given in the previous section we know that there exists a square integrable degenerate kernel (45) such that

$$K(x, s) = P(x, s) + Q(x, s) \tag{47}$$

and

$$\|\boldsymbol{Q}\|_2 < \frac{1}{\omega}. \tag{48}$$

Such a *dissection* of the kernel $K(x, s)$ was introduced by Schmidt in 1907 and it enables us to write the integral equation (46) as

$$\phi(x) = f_1(x; \lambda) + \lambda \int_a^b Q(x, s)\phi(s)\, ds \tag{49}$$

where

$$f_1(x; \lambda) = f(x) + \lambda \int_a^b P(x, s)\phi(s)\, ds. \tag{50}$$

Since $|\lambda| \|\boldsymbol{Q}\|_2 < 1$ it follows that the Neumann series for the resolvent kernel $R_Q(x, s; \lambda)$ of $Q(x, s)$ is convergent, so that the solution of the integral equation (49) can be expressed in the form

$$\phi(x) = f_1(x; \lambda) + \lambda \int_a^b R_Q(x, s; \lambda)f_1(s; \lambda)\, ds. \tag{51}$$

This may be rewritten

$$\phi(x) = f_2(x; \lambda) + \lambda \int_a^b T(x, s; \lambda)\phi(s)\, ds \tag{52}$$

where

$$f_2(x; \lambda) = f(x) + \lambda \int_a^b R_Q(x, s; \lambda)f(s)\, ds \tag{53}$$

and

$$T(x, s; \lambda) = P(x, s) + \lambda \int_a^b R_Q(x, t; \lambda)P(t, s)\, dt. \tag{54}$$

Here $T(x, s; \lambda)$ is a degenerate kernel since it can be expressed in the form

$$T(x, s; \lambda) = \sum_{i=1}^n y_i(x)\overline{v_i(s)} \tag{55}$$

where

$$y_i(x) = u_i(x) + \lambda \int_a^b R_Q(x, t; \lambda) u_i(t) \, \mathrm{d}t. \tag{56}$$

Thus we have converted our original integral equation (46) possessing a general square integrable kernel $K(x, s)$ into an integral equation (52) with a degenerate kernel $T(x, s; \lambda)$. In particular we note that if our original equation is homogeneous so that $f(x) = 0$, the new equation with degenerate kernel $T(x, s; \lambda)$ is also homogeneous since $f_2(x; \lambda) = 0$. This means that the results obtained in section 9.1 on degenerate kernels can now be applied to the present case of a general kernel.

Using the expression (12) for the resolvent kernel corresponding to a degenerate kernel we obtain

$$\phi(x) = f_2(x, \lambda) + \frac{\lambda}{d(\lambda)} \int_a^b \sum_{i=1}^n \sum_{j=1}^n y_i(x) d_{ij}(\lambda) \overline{v_j(s)} f_2(s; \lambda) \, \mathrm{d}s \tag{57}$$

where $d(\lambda) = \det (\mathbf{I} - \lambda \mathbf{A})$ and the matrix $\mathbf{A}$ has elements $a_{ij} = \int_a^b \overline{v_i(s)} y_j(s) \, \mathrm{d}s$ while the $d_{ij}(\lambda)$ are the elements of the adjugate matrix $\mathbf{D}$ of $\mathbf{I} - \lambda \mathbf{A}$. Hence the resolvent kernel of $K(x, s)$ can be seen to have the form

$$R(x, s; \lambda) = R_Q(x, s; \lambda) + \frac{1}{d(\lambda)} \sum_{i=1}^n \sum_{j=1}^n y_i(x) d_{ij}(\lambda) \overline{z_j(s)} \tag{58}$$

where

$$z_j(s) = v_j(s) + \bar{\lambda} \int_a^b v_j(t) \overline{R_Q(t, s; \lambda)} \, \mathrm{d}t. \tag{59}$$

The resolvent kernel $R(x, s; \lambda)$ is therefore an analytic function of λ for $|\lambda| < \omega$ except for poles occurring at the zeros of $d(\lambda)$. For a given ω these poles are finite in number and so as $\omega \to \infty$ it is clear that the number of poles becomes at most denumerably infinite with no finite points of accumulation. This means that $R(x, s; \lambda)$ is a meromorphic function of λ whose poles coincide with the characteristic values of $\mathbf{K}$.

Our results can be expressed in the form of a set of theorems known as the *Fredholm alternatives*:

Theorem 1. If $\boldsymbol{K}$ is an integral operator with a square integrable kernel $K(x, s)$, and

$$\phi = f + \lambda \boldsymbol{K} \phi \tag{60}$$

is a Fredholm integral equation of the second kind, then λ is either a regular value or a characteristic value of $\boldsymbol{K}$.

If λ is a regular value of $\boldsymbol{K}$, the integral equation (60) has the unique square integrable solution

$$\phi = f + \lambda \boldsymbol{R} f \tag{61}$$

for any square integrable function f, where the resolvent operator $\boldsymbol{R}$ with kernel $R(x, s; \lambda)$ is a meromorphic function of λ.

The characteristic values of $\boldsymbol{K}$ are at most denumerably infinite with no finite accumulation point. For each characteristic value λ of rank p, the homogeneous equation

$$\phi = \lambda \boldsymbol{K} \phi \tag{62}$$

has p linearly independent solutions $\phi^{(k)}$ $(k = 1, \ldots, p)$.

Theorem 2. If λ is a characteristic value of $\boldsymbol{K}$ then $\bar{\lambda}$ is a characteristic value of $\boldsymbol{K}^*$. The number of linearly independent solutions of (62) and the adjoint homogeneous equation

$$\psi = \bar{\lambda} \boldsymbol{K}^* \psi \tag{63}$$

are the same.

Theorem 3. If λ is a characteristic value of $\boldsymbol{K}$ and f is a square integrable function, the integral equation (60) has a square integrable solution if and only if f is orthogonal to every square integrable solution of the adjoint homogeneous equation (63). Then the general square integrable solution of (60) is given by

$$\phi = \phi_0 + \sum_{k=1}^{p} \alpha_k \phi^{(k)} \tag{64}$$

where ϕ_0 is a particular square integrable solution of (60).

9.3.1 Fredholm theorems for completely continuous operators

The Fredholm theorems stated above were generalized by Hilbert to hold for completely continuous operators.

As for integral operators, the demonstration depends upon the introduction of a degenerate, that is finite dimensional, linear operator P.

Suppose K is a bounded linear operator and let $v_1, v_2, \ldots, v_i, \ldots$ be a complete orthonormal system in abstract Hilbert space H. Then the kernel matrix of K has elements

$$k_{ij} = (Kv_i, v_j) \tag{65}$$

Putting

$$Kv_i = u_i \quad (i = 1, 2, \ldots) \tag{66}$$

and expanding u_i in the form

$$u_i = \sum_{j=1}^{\infty} x_j^{(i)} v_j \tag{67}$$

where

$$\|u_i\|^2 = \sum_{j=1}^{\infty} |x_j^{(i)}|^2 < \infty, \tag{68}$$

we have

$$\sum_{i=1}^{\infty} \sum_{j=1}^{\infty} |k_{ij}|^2 = \sum_{i=1}^{\infty} \sum_{j=1}^{\infty} |(u_i, v_j)|^2 = \sum_{i=1}^{\infty} \sum_{j=1}^{\infty} |x_j^{(i)}|^2$$

$$= \sum_{i=1}^{\infty} \|u_i\|^2. \tag{69}$$

Let us now assume that

$$\sum_{i=1}^{\infty} \sum_{j=1}^{\infty} |k_{ij}|^2 < \infty \tag{70}$$

so that K is completely continuous as shown in section 7.3. Then it follows from (69) that, given any $\varepsilon > 0$, there exists N such that if $n > N$

$$\sum_{i=n+1}^{\infty} \|u_i\|^2 < \varepsilon^2. \tag{71}$$

Next we introduce a degenerate operator $\boldsymbol{P}$ such that

$$\begin{aligned}\boldsymbol{P}v_i &= u_i \qquad (i=1,\ldots,n)\\ &= 0 \qquad (i>n).\end{aligned} \tag{72}$$

Suppose that v is any element in H and

$$v = \sum_{i=1}^{\infty} y_i v_i \tag{73}$$

where

$$\|v\|^2 = \sum_{i=1}^{\infty} |y_i|^2 < \infty. \tag{74}$$

Then we have

$$\begin{aligned}\|(\boldsymbol{K}-\boldsymbol{P})v\| &= \left\| \sum_{i=1}^{\infty} y_i (\boldsymbol{K}-\boldsymbol{P})v_i \right\|\\ &\leqslant \sum_{i=1}^{\infty} |y_i|\,\|(\boldsymbol{K}-\boldsymbol{P})v_i\|\\ &= \sum_{i=n+1}^{\infty} |y_i|\,\|u_i\|\\ &\leqslant \sqrt{\left(\sum_{i=n+1}^{\infty} |y_i|^2 \right)\left(\sum_{i=n+1}^{\infty} \|u_i\|^2 \right)}\\ &< \|v\|\varepsilon\end{aligned} \tag{75}$$

using the triangle inequality and Cauchy's inequality together with (71) and (74). Thus, by choosing n sufficiently large, we can ensure that $\|\boldsymbol{K}-\boldsymbol{P}\| < \varepsilon$ provided the kernel matrix of $\boldsymbol{K}$ satisfies (70) so that $\boldsymbol{K}$ is completely continuous. This result allows us to approximate a completely continuous operator $\boldsymbol{K}$ as closely as we please by a suitable degenerate operator $\boldsymbol{P}$. Setting $\boldsymbol{Q} = \boldsymbol{K} - \boldsymbol{P}$ and choosing n so large that $|\lambda|\,\|\boldsymbol{Q}\| < 1$, the resolvent operator $\boldsymbol{R}_Q$ of $\boldsymbol{Q}$ is bounded and given by a convergent Neumann series, as shown in section 8.5, which enables us to carry through a similar analysis to that presented in section 9.3.

9.4   Fredholm formulae for continuous kernels

To conclude this chapter we shall derive the Fredholm solution of the integral equation

$$\phi(x) = f(x) + \lambda \int_a^b K(x,s)\phi(s)\,\mathrm{d}s \,(a \leqslant x \leqslant b) \tag{76}$$

assuming that the kernel $K(x, s)$ is continuous in the square domain $a \leq x \leq b$, $a \leq s \leq b$, that $f(x)$ is continuous in the interval $a \leq x \leq b$, and that the integration is performed in the Riemann sense.

We first note the solution of (76) must be continuous for we have

$$|\phi(x) - \phi(x')| \leq |f(x) - f(x')| |\lambda| \left\{ \int_a^b |K(x, s) - K(x', s)|^2 \, ds \right\}^{1/2}$$

using the triangle and Schwarz inequalities, so that the continuity of $K(x, s)$ and $f(x)$ ensures the continuity of the square integrable solution $\phi(x)$.

We shall see that the solution obtained by Fredholm involves a resolvent kernel given by the ratio of two power series in λ which are convergent for all λ. The method introduced by Fredholm in 1903 to establish his formulae treats the integral equation (76) as the limiting form of a finite system of linear algebraic equations. These equations are obtained by choosing a net of n equal sub-intervals having length $\delta_n = (b - a)/n$ given by

$$a = x_0 < x_1 < x_2 < \ldots < x_r < \ldots < x_n = b \tag{77}$$

where $x_r = a + r\delta_n$. Approximating the Riemann integral on the right-hand side of (76) by the finite sum

$$\delta_n \sum_{s=1}^n K(x_r, x_s) \phi(x_s)$$

where $x = x_r$, we may replace the integral equation by the system of n algebraic equations

$$\phi(x_r) - \lambda \delta_n \sum_{s=1}^n K(x_r, x_s) \phi(x_s) = f(x_r). \tag{78}$$

Provided the matrix

$$I - \lambda \delta_n K =$$

$$\begin{pmatrix}
1 - \lambda \delta_n K(x_1, x_1) & -\lambda \delta_n K(x_1, x_2) & \ldots & -\lambda \delta_n K(x_1, x_n) \\
-\lambda \delta_n K(x_2, x_1) & 1 - \lambda \delta_n K(x_2, x_2) & \ldots & -\lambda \delta_n K(x_2, x_n) \\
\cdot & \cdot & & \cdot \\
\cdot & \cdot & & \cdot \\
\cdot & \cdot & & \cdot \\
-\lambda \delta_n K(x_n, x_1) & -\lambda \delta_n K(x_n, x_2) & \ldots & 1 - \lambda \delta_n K(x_n, x_n)
\end{pmatrix}$$

$$\tag{79}$$

has the non-vanishing determinant $d_n(\lambda) = \det(\mathbf{I} - \lambda \delta_n \mathbf{K})$, the system of equations has the unique solution

$$\phi(x_r) = \frac{1}{d_n(\lambda)} \sum_{s=1}^{n} D_n(x_r, x_s) f(x_s) \qquad (80)$$

where $D_n(x_r, x_s)$ is the r, s element of the adjugate matrix of $\mathbf{I} - \lambda \delta_n \mathbf{K}$, that is $D_n(x_r, x_s)$ is the cofactor of the element involving $K(x_s, x_r)$ in $\mathbf{I} - \lambda \delta_n \mathbf{K}$.

Expanding $d_n(\lambda)$ in powers of λ we obtain

$$d_n(\lambda) = 1 - \frac{\lambda \delta_n}{1!} S_1 + \frac{\lambda^2 \delta_n^2}{2!} S_2 - \ldots + \frac{(-1)^n \lambda^n \delta_n^n}{n!} S_n \qquad (81)$$

where

$$S_1 = \sum_{r=1}^{n} K(x_r, x_r),$$

$$S_2 = \sum_{r_1, r_2 = 1}^{n} \begin{vmatrix} K(x_{r_1}, x_{r_1}) & K(x_{r_1}, x_{r_2}) \\ K(x_{r_2}, x_{r_1}) & K(x_{r_2}, x_{r_2}) \end{vmatrix},$$

and generally

$$S_m = \sum_{r_1, r_2, \ldots, r_m = 1}^{n} \begin{vmatrix} K(x_{r_1}, x_{r_1}) & K(x_{r_1}, x_{r_2}) & \ldots & K(x_{r_1}, x_{r_m}) \\ K(x_{r_2}, x_{r_1}) & K(x_{r_2}, x_{r_2}) & \ldots & K(x_{r_2}, x_{r_m}) \\ \cdot & \cdot & & \cdot \\ \cdot & \cdot & & \cdot \\ \cdot & \cdot & & \cdot \\ K(x_{r_m}, x_{r_1}) & K(x_{r_m}, x_{r_2}) & \ldots & K(x_{r_m}, x_{r_m}) \end{vmatrix} \qquad (82)$$

Now letting $n \to \infty$ and $\delta_n \to 0$ we obtain $d_n(\lambda) \to d(\lambda)$ where

$$d(\lambda) = 1 - \lambda \int_a^b K(x_1, x_1) \, dx_1 + \frac{\lambda^2}{2!} \int_a^b \int_a^b \begin{vmatrix} K(x_1, x_1) & K(x_1, x_2) \\ K(x_2, x_1) & K(x_2, x_2) \end{vmatrix} \, dx_1 \, dx_2$$

$$- \frac{\lambda^3}{3!} \int_a^b \int_a^b \int_a^b \begin{vmatrix} K(x_1, x_1) & K(x_1, x_2) & K(x_1, x_3) \\ K(x_2, x_1) & K(x_2, x_2) & K(x_2, x_3) \\ K(x_3, x_1) & K(x_3, x_2) & K(x_3, x_3) \end{vmatrix} \, dx_1 \, dx_2 \, dx_3$$

$$+ \ldots \qquad (83)$$

Introducing the notation

$$K\begin{pmatrix} x_1, x_2, \ldots, x_m \\ s_1, s_2, \ldots, s_m \end{pmatrix} =$$

$$\begin{vmatrix} K(x_1, s_1) & K(x_1, s_2) & \ldots & K(x_1, s_m) \\ K(x_2, s_1) & K(x_2, s_2) & \ldots & K(x_2, s_m) \\ \cdot & \cdot & & \cdot \\ \cdot & \cdot & & \cdot \\ \cdot & \cdot & & \cdot \\ K(x_m, s_1) & K(x_m, s_2) & \ldots & K(x_m, s_m) \end{vmatrix}$$

(84)

we see that

$$d(\lambda) = \sum_{m=0}^{\infty} d_m \lambda^m \qquad (85)$$

where $d_0 = 1$ and

$$d_m = \frac{(-1)^m}{m!} \int_a^b \int_a^b \ldots \int_a^b K\begin{pmatrix} x_1, x_2, \ldots, x_m \\ x_1, x_2, \ldots, x_m \end{pmatrix} dx_1 \, dx_2 \ldots dx_m. \quad (86)$$

$d(\lambda)$ is known as the *Fredholm determinant*.

Now, $D_n(x_r, x_s)$ being the cofactor of the (s, r) element in $d_n(\lambda)$, we have for $r \neq s$

$$D_n(x_r, x_s) = \lambda \delta_n \left[K(x_r, x_s) - \lambda \delta_n \sum_{s_1=1}^{n} \begin{vmatrix} K(x_r, x_s) & K(x_r, x_{s_1}) \\ K(x_{s_1}, x_s) & K(x_{s_1}, x_{s_1}) \end{vmatrix} \right.$$

$$\left. + \frac{\lambda^2 \delta_n^2}{2!} \sum_{s_1, s_2=1}^{n} \begin{vmatrix} K(x_r, x_s) & K(x_r, x_{s_1}) & K(x_r, x_{s_2}) \\ K(x_{s_1}, x_s) & K(x_{s_1}, x_{s_1}) & K(x_{s_1}, x_{s_2}) \\ K(x_{s_2}, x_s) & K(x_{s_2}, x_{s_1}) & K(x_{s_2}, x_{s_2}) \end{vmatrix} - \ldots \right]$$

(87)

and so, letting $n \to \infty$ and $x_r \to x$, $x_s \to s$, we see that $\delta_n^{-1} D_n(x_r, x_s) \to \lambda D(x, s; \lambda)$ where

$$D(x, s; \lambda) = K(x, s) - \lambda \int_a^b \begin{vmatrix} K(x, s) & K(x, s_1) \\ K(s_1, s) & K(s_1, s_1) \end{vmatrix} ds_1$$

$$+ \frac{\lambda^2}{2!} \int_a^b \int_a^b \begin{vmatrix} K(x, s) & K(x, s_1) & K(x, s_2) \\ K(s_1, s) & K(s_1, s_1) & K(s_1, s_2) \\ K(s_2, s) & K(s_2, s_1) & K(s_2, s_2) \end{vmatrix} ds_1 \, ds_2 - \ldots$$

(88)

However for $r = s$, D_n has a similar expansion to $d_n(\lambda)$ and in fact it can be readily verified that $D_n(x_r, x_r) \to d(\lambda)$ as $n \to \infty$. Hence in the limit as $n \to \infty$ we obtain

$$\phi(x) = f(x) + \lambda \int_a^b R(x, s; \lambda) f(s) \, ds \qquad (89)$$

where the resolvent kernel $R(x, s; \lambda)$ is given by

$$R(x, s; \lambda) = \frac{D(x, s; \lambda)}{d(\lambda)} \qquad (90)$$

with

$$D(x, s; \lambda) = \sum_{m=0}^{\infty} D_m(x, s) \lambda^m \qquad (91)$$

and $D_0(x, s) = K(x, s)$ while generally

$$D_m(x, s) = \frac{(-1)^m}{m!} \int_a^b \int_a^b \dots \int_a^b K\begin{pmatrix} x, s_1, s_2, \dots, s_m \\ s, s_1, s_2, \dots, s_m \end{pmatrix} ds_1 \, ds_2 \dots ds_m. \qquad (92)$$

$D(x, s; \lambda)$ is known as the *first Fredholm minor*. To establish the convergence of the series for $d(\lambda)$ and $D(x, s; \lambda)$ we employ *Hadamard's inequality*

$$|\det \mathbf{A}|^2 \le \prod_{r=1}^n \sum_{s=1}^n |a_{rs}|^2 \qquad (93)$$

where $\mathbf{A}$ is a $n \times n$ matrix whose (r, s) element is the complex number a_{rs}.

This inequality has a simple interpretation in three-dimensional real Euclidean space. If $\mathbf{a}_1 = (a_{11}, a_{12}, a_{13})$, $\mathbf{a}_2 = (a_{21}, a_{22}, a_{23})$, $\mathbf{a}_3 = (a_{31}, a_{32}, a_{33})$ are three vectors determining the sides of a parallelepiped, its volume is given by the scalar triple product

$$\mathbf{a}_1 \cdot (\mathbf{a}_2 \times \mathbf{a}_3) = \begin{vmatrix} a_{11} & a_{12} & a_{13} \\ a_{21} & a_{22} & a_{23} \\ a_{31} & a_{32} & a_{33} \end{vmatrix}.$$

This is less than or equal to the volume of the corresponding rectangular parallelepiped which is $\|\mathbf{a}_1\| \|\mathbf{a}_2\| \|\mathbf{a}_3\|$ and thus (93) follows for $n = 3$ and real numbers a_{rs}.

If $|a_{rs}| \le M$ for all r, s we have

$$|\det \mathbf{A}|^2 \le n^n M^{2n}. \qquad (94)$$

Since the kernel $K(x, s)$ is continuous it is also bounded and so there exists a positive number M such that $|K(x, s)| \le M$. Hence

$$|d_m| \le \frac{m^{m/2} M^m (b-a)^m}{m!} = a_m \tag{95}$$

and so

$$|d(\lambda)| \le \sum_{m=0}^{\infty} |d_m| |\lambda|^m \le \sum_{m=0}^{\infty} a_m |\lambda|^m. \tag{96}$$

Now

$$\frac{a_{m+1}}{a_m} = \frac{\left(1+\dfrac{1}{m}\right)^{m/2} M(b-a)}{(m+1)^{1/2}} \to 0 \tag{97}$$

as $m \to \infty$ since $(1+1/m)^m \to e$ and so $\sum_{m=0}^{\infty} a_m |\lambda|^m$ is convergent for all λ. Hence (96) implies the absolute convergence of $d(\lambda)$ for all λ.

Also, by the inequality (94), we have

$$|D_m(x, s)| \le \frac{(m+1)^{(m+1)/2} M^{m+1} (b-a)^m}{m!} = b_m \tag{98}$$

and so

$$|D(x, s; \lambda)| \le \sum_{m=0}^{\infty} |D_m(x, s)| |\lambda|^m \le \sum_{m=0}^{\infty} b_m |\lambda|^m. \tag{99}$$

Now

$$\frac{b_m}{b_{m-1}} = \frac{(m+1)^{1/2}}{m} \left(1+\frac{1}{m}\right)^{m/2} M(b-a) \to 0 \tag{100}$$

as $m \to \infty$, from which it follows that $\sum_{m=0}^{\infty} b_m |\lambda|^m$ is convergent so that $D(x, s; \lambda)$ is absolutely and uniformly convergent for all λ.

Our next step is to verify that the Fredholm solution (89) is indeed correct by showing that the resolvent kernel (90) satisfies the resolvent equation. To this end we note that

$$D_m(x, s) = d_m K(x, s) + Q_m(x, s) \tag{101}$$

where

$$Q_m(x, s) = \frac{(-1)^m}{m!} \int_a^b \dots \int_a^b
\begin{vmatrix}
0 & K(x, s_1) & \dots & K(x, s_m) \\
K(s_1, s) & K(s_1, s_1) & \dots & K(s_1, s_m) \\
\cdot & \cdot & & \cdot \\
\cdot & \cdot & & \cdot \\
\cdot & \cdot & & \cdot \\
K(s_m, s) & K(s_m, s_1) & & K(s_m, s_m)
\end{vmatrix} ds_1 \dots ds_m \tag{102}$$

If we now expand in terms of the minors of the first column we obtain

$$Q_m(x, s) = \frac{(-1)^m}{m!} \sum_{i=1}^m (-1)^i \int_a^b \cdots \int_a^b K\begin{pmatrix} x, s_1, \ldots, s_{i-1}, s_{i+1}, \ldots, s_m \\ s_1, s_2, \ldots, s_i, s_{i+1}, \ldots, s_m \end{pmatrix}$$
$$\times K(s_i, s) \, ds_1 \ldots ds_m$$

$$= \frac{(-1)^{m-1}}{(m-1)!} \int_a^b \cdots \int_a^b \int_a^b K\begin{pmatrix} x, s_1, \ldots, s_{m-1} \\ t, s_1, \ldots, s_{m-1} \end{pmatrix} K(t, s) \, ds_1 \ldots ds_{m-1} \, dt$$

$$= \int_a^b D_{m-1}(x, t) K(t, s) \, dt \tag{103}$$

which yields the recurrence relation

$$D_m(x, s) = d_m K(x, s) + \int_a^b D_{m-1}(x, t) K(t, s) \, dt. \tag{104}$$

Hence

$$D(x, s; \lambda) = d(\lambda) K(x, s) + \lambda \int_a^b D(x, t; \lambda) K(t, s) \, dt. \tag{105}$$

Similarly by expanding in terms of the minors of the first row of (102), we can show that

$$D_m(x, s) = d_m K(x, s) + \int_a^b K(x, t) D_{m-1}(t, s) \, dt \tag{106}$$

and so we have also

$$D(x, s; \lambda) = d(\lambda) K(x, s) + \lambda \int_a^b K(x, t) D(t, s; \lambda) \, dt. \tag{107}$$

It follows that the resolvent kernel (90) satisfies the resolvent equation

$$R(x, s; \lambda) - K(x, s) = \lambda \int_a^b R(x, t; \lambda) K(t, s) \, dt$$

$$= \lambda \int_a^b K(x, t) R(t, s; \lambda) \, dt \tag{108}$$

and thus the Fredholm solution is verified.

Since

$$D_{m-1}(x, s) = \frac{(-1)^{m-1}}{(m-1)!} \int_a^b \cdots \int_a^b K\begin{pmatrix} x, s_1, \ldots, s_{m-1} \\ s, s_1, \ldots, s_{m-1} \end{pmatrix} ds_1 \ldots ds_{m-1}$$

it follows that

$$\int_a^b D_{m-1}(s, s)\, ds = \frac{(-1)^{m-1}}{(m-1)!} \int_a^b \int_a^b \cdots \int_a^b K\begin{pmatrix} s, s_1, \ldots, s_{m-1} \\ s, s_1, \ldots, s_{m-1} \end{pmatrix} ds\, ds_1 \\ \ldots ds_{m-1}$$

and so we obtain the relation

$$d_m = -\frac{1}{m} \int_a^b D_{m-1}(s, s)\, ds. \tag{109}$$

Also, using (109), we have that

$$d'(\lambda) = \sum_{m=1}^{\infty} m d_m \lambda^{m-1}$$

$$= -\sum_{m=0}^{\infty} \lambda^m \int_a^b D_m(s, s)\, ds$$

$$= -\int_a^b D(s, s; \lambda)\, ds \tag{110}$$

and so

$$\frac{d'(\lambda)}{d(\lambda)} = -\int_a^b R(s, s; \lambda)\, ds$$

$$= -\sum_{n=0}^{\infty} \lambda^n \int_a^b K_{n+1}(s, s)\, ds \tag{111}$$

employing the Neumann expansion (8.22) for the resolvent kernel.

The *trace* of the kernel $K(x, s)$ is defined to be the quantity

$$\kappa_1 = \int_a^b K(s, s)\, ds \tag{112}$$

while the trace of the iterate $K_n(x, s)$ is given by

$$\kappa_n = \int_a^b \cdots \int_a^b \int_a^b K(s, t_1) K(t_1, t_2) \ldots K(t_{n-1}, s)\, dt_1 \ldots dt_{n-1}\, ds. \tag{113}$$

Hence we obtain the formula

$$\frac{d'(\lambda)}{d(\lambda)} = -\sum_{n=0}^{\infty} \kappa_{n+1} \lambda^n \tag{114}$$

from which it follows that the radius of convergence of the power series on the right-hand side is $|\lambda_1|$ where λ_1 is the characteristic value of the kernel $K(x, s)$ having the least absolute magnitude.

Example. To illustrate the application of the Fredholm recurrence formulae (104) and (109) obtained above we consider the kernel

$$K(x, s) = x + s \quad (0 \le x \le 1, 0 \le s \le 1). \tag{115}$$

We have

$$d_0 = 1, D_0(x, s) = K(x, s) = x + s,$$

$$d_1 = -\int_0^1 D_0(s, s)\, \mathrm{d}s = -\int_0^1 2s\, \mathrm{d}s = -1$$

and

$$D_1(x, s) = -K(x, s) + \int_0^1 D_0(x, t)K(t, s)\, \mathrm{d}t$$

$$= -(x + s) + \int_0^1 (x + t)(t + s)\, \mathrm{d}t$$

$$= \frac{1}{3} - \frac{1}{2}(x + s) + xs.$$

Hence

$$d_2 = -\frac{1}{2}\int_0^1 D_1(s, s)\, \mathrm{d}s.$$

$$= -\frac{1}{2}\int_0^1 (\tfrac{1}{3} - s + s^2)\, \mathrm{d}s$$

$$= -\tfrac{1}{12}.$$

Further

$$D_2(x, s) = -\tfrac{1}{12}K(x, s) + \int_0^1 D_1(x, t)K(t, s)\, \mathrm{d}t = 0$$

from which it follows that $d_3 = 0$. Repeated application of the recurrence relations leads to

$$D_m(x, s) = 0 \quad (m \geqslant 2), \quad d_m = 0 \quad (m \geqslant 3).$$

Thus

$$d(\lambda) = 1 - \lambda - \tfrac{1}{12}\lambda^2$$

and

$$D(x, s; \lambda) = x + s + \{\tfrac{1}{3} - \tfrac{1}{2}(x + s) + xs\}\lambda.$$

Problems

1. Obtain the resolvent kernels for the symmetric kernels
   (i) $K(x, s) = 1 + xs$ $(0 \leqslant x \leqslant 1, 0 \leqslant s \leqslant 1)$,
   (ii) $K(x, s) = x^2 + s^2$ $(0 \leqslant x \leqslant 1, 0 \leqslant s \leqslant 1)$,
   (iii) $K(x, s) = xs + x^2 s^2$ $(-1 \leqslant x \leqslant 1, -1 \leqslant s \leqslant 1)$.
   Verify that their characteristic values are real numbers.

2. Obtain the resolvent kernels for the skew-symmetric kernels
   (i) $K(x, s) = \sin(x - s)$ $(0 \leqslant x \leqslant 2\pi, 0 \leqslant s \leqslant 2\pi)$,
   (ii) $K(x, s) = x^2 - s^2$ $(0 \leqslant x \leqslant 1, 0 \leqslant s \leqslant 1)$,
   (iii) $K(x, s) = x^2 s - xs^2$ $(0 \leqslant x \leqslant 1, 0 \leqslant s \leqslant 1)$.
   Verify that their characteristic values are pure imaginary numbers.

3. Show that the non symmetric kernels
   (i) $K(x, s) = \sin x \cos s$ $(0 \leqslant x \leqslant \pi, 0 \leqslant s \leqslant \pi)$,
   (ii) $K(x, s) = \sin x \sin 2s$ $(0 \leqslant x \leqslant \pi, 0 \leqslant s \leqslant \pi)$,
   have no characteristic values.

4. Use the theory on degenerate kernels given in section 9.1 to solve the Fredholm integral equations of the second kind possessing the degenerate kernels
   (i) $K(x, s) = x + s$ $(0 \leqslant x \leqslant 1, 0 \leqslant s \leqslant 1)$,
   (ii) $K(x, s) = \cos(x + s)$ $(0 \leqslant x \leqslant 2\pi, 0 \leqslant s \leqslant 2\pi)$.

5. Use the Fredholm formulae given in section 9.4 to solve the Fredholm integral equations of the second kind possessing the kernels
   (i) $K(x, s) = \sin(x + s)$ $(0 \leqslant x \leqslant 2\pi, 0 \leqslant s \leqslant 2\pi)$,
   (ii) $K(x, s) = x - s$ $(0 \leqslant x \leqslant 1, 0 \leqslant s \leqslant 1)$,
   (iii) $K(x, s) = 1 - 3xs$ $(0 \leqslant x \leqslant 1, 0 \leqslant s \leqslant 1)$.
   Verify that your solutions to (i) and (ii) agree with examples 1 and 2 of section 9.1.

10

Hilbert-Schmidt theory

In this final chapter we shall direct our attention to self-adjoint or Hermitian integral operators K satisfying

$$(K\phi, \psi) = (\phi, K\psi) \tag{1}$$

for all ϕ, ψ belonging to the Hilbert space of L^2 functions. Our objective will be to develop the theory originated by Hilbert and Schmidt, in which the resolvent kernel $R(x, s; \lambda)$ was expanded in terms of the characteristic functions $\phi_\nu(x)$ and characteristic values λ_ν of the Hermitian kernel $K(x, s)$.

10.1 Hermitian kernels

The original theory of Hilbert and Schmidt dealt with real symmetric kernels satisfying the condition

$$K(s, x) = K(x, s)$$

but we shall be studying the more general case of Hermitian kernels for which

$$\overline{K(s, x)} = K(x, s) \tag{2}$$

where $K(x, s)$ is not necessarily real. Then the integral operator

$$K = \int_a^b K(x, s) \, \mathrm{d}s \tag{3}$$

is Hermitian since it satisfies the relation (1).

We see from (1) that $(K\psi, \psi) = (\psi, K\psi) = \overline{(K\psi, \psi)}$ and so $(K\psi, \psi)$ is real for all L^2 functions ψ.

A Hermitian kernel which is square integrable is called a *Hilbert-Schmidt* kernel.

10.2 Spectrum of a Hilbert-Schmidt kernel

Let us begin by investigating the properties of the set of characteristic values or *spectrum* of a Hermitian square integrable kernel, that is a Hilbert-Schmidt kernel $K(x, s)$.

We shall show first that every non-null operator $\boldsymbol{K}$ possessing a Hilbert-Schmidt kernel has at least one characteristic value λ_1. The method of proof is due to Kneser.

Since $\boldsymbol{K}$ is Hermitian it follows from (1) that $\boldsymbol{K}^n$ is also Hermitian and hence, using the definition (9.112) of the trace of an integral operator with kernel $K(x, s)$, we have

$$
\begin{aligned}
\kappa_{2n} &= \text{trace } (\boldsymbol{K}^n \boldsymbol{K}^n) \\
&= \int_a^b \int_a^b K_n(x, s) K_n(s, x) \, ds \, dx \\
&= \int_a^b \int_a^b K_n(x, s) \overline{K_n(x, s)} \, ds \, dx \\
&= \|\boldsymbol{K}^n\|_2^2
\end{aligned}
\tag{4}
$$

and so $\kappa_{2n} \geq 0$. Also $\kappa_{2n} < \infty$ since $\|\boldsymbol{K}^n\|_2 \leq \|\boldsymbol{K}\|_2^n$ and $K(x, s)$ is square integrable..

Using Schwarz's inequality for double integrals we have

$$
\begin{aligned}
\kappa_{2n}^2 &= \{\text{trace } (\boldsymbol{K}^{n-1} \boldsymbol{K}^{n+1})\}^2 \\
&= \left\{ \int_a^b \int_a^b K_{n-1}(x, s) K_{n+1}(s, x) \, ds \, dx \right\}^2 \\
&\leq \left\{ \int_a^b \int_a^b |K_{n-1}(x, s)|^2 \, ds \, dx \right\} \left\{ \int_a^b \int_a^b |K_{n+1}(x, s)|^2 \, ds \, dx \right\} \\
&= \kappa_{2n-2} \kappa_{2n+2} \qquad (n \geq 2).
\end{aligned}
\tag{5}
$$

Now $\|\boldsymbol{K}\|_2 > 0$ since $\boldsymbol{K}$ is non-null and so $\kappa_2 > 0$. Further $\kappa_4 > 0$, for if we were to suppose that $\kappa_4 = 0$ it would imply that $K_2(x, s) = 0$ almost everywhere and in particular we would have $K_2(s, s) = 0$ so that $\kappa_2 = 0$. Since $\kappa_2 > 0$ and $\kappa_4 > 0$ it follows from (5) that $\kappa_{2n} > 0$ for all n and

$$
\frac{\kappa_{2n+2}}{\kappa_{2n}} \geq \frac{\kappa_{2n}}{\kappa_{2n-2}} \geq \ldots \geq \frac{\kappa_4}{\kappa_2}.
\tag{6}
$$

Putting $u_n = \kappa_{2n} |\lambda|^{2n-1}$ we see that

$$
\frac{u_{n+1}}{u_n} = |\lambda|^2 \frac{\kappa_{2n+2}}{\kappa_{2n}} \geq |\lambda|^2 \frac{\kappa_4}{\kappa_2}
$$

and hence $\sum_{n=1}^{\infty} u_n$ is divergent if $|\lambda| > \sqrt{\kappa_2/\kappa_4}$. But $\sum_{n=0}^{\infty} \kappa_{n+1} \lambda^n$ is convergent for sufficiently small $|\lambda|$ and indeed we showed at the end of section 9.4 that the radius of convergence of this series is $|\lambda_1|$

where λ_1 is the characteristic value of $\boldsymbol{K}$ having least absolute magnitude. Hence $\sum_{n=1}^{\infty} u_n$ is also convergent for $|\lambda| < |\lambda_1|$ and since we already know that this series is divergent for $|\lambda| > \sqrt{\kappa_2/\kappa_4}$, it follows that there exists at least one characteristic value λ_1 and that $|\lambda_1| \leqslant \sqrt{\kappa_2/\kappa_4}$.

Next we shall show that all the characteristic values of a Hermitian kernel $K(x, s)$ are real. Thus let

$$\lambda \boldsymbol{K} \phi = \phi$$

where ϕ is a non-null characteristic function associated with the characteristic value λ. Then

$$(\boldsymbol{K}\phi, \phi) = (\lambda^{-1}\phi, \phi) = \lambda^{-1}(\phi, \phi)$$

and since $(\boldsymbol{K}\phi, \phi)$ is real, and (ϕ, ϕ) is real and non-zero, it follows that λ is real.

We also see that the characteristic functions ϕ_1, ϕ_2 of a Hermitian kernel $K(x, s)$ for different characteristic values λ_1, λ_2 are orthogonal. For we have

$$(\phi_1, \phi_2) = (\lambda_1 \boldsymbol{K}\phi_1, \phi_2) = \lambda_1(\boldsymbol{K}\phi_1, \phi_2)$$

and

$$(\phi_1, \phi_2) = (\phi_1, \lambda_2 \boldsymbol{K}\phi_2) = \lambda_2(\boldsymbol{K}\phi_1, \phi_2)$$

since $\boldsymbol{K}$ is Hermitian and λ_2 is real. This yields

$$(\lambda_1^{-1} - \lambda_2^{-1})(\phi_1, \phi_2) = 0$$

and as $\lambda_1 \neq \lambda_2$ it follows that $(\phi_1, \phi_2) = 0$.

We proved above that every non-null Hermitian kernel $K(x, s)$ possesses at least one characteristic value λ_1 and associated normalized characteristic function ϕ_1. Now consider the Hermitian kernel

$$K^{(2)}(x, s) = K(x, s) - \frac{\phi_1(x)\overline{\phi_1(s)}}{\lambda_1}. \tag{7}$$

If $K^{(2)}(x, s)$ is non-null it also must have at least one characteristic value λ_2 and associated characteristic function ϕ_2.

Since

$$\int_a^b K^{(2)}(x, s)\phi_1(s)\,\mathrm{d}s = \int_a^b K(x, s)\phi_1(s)\,\mathrm{d}s - \frac{\phi_1(x)}{\lambda_1} = 0$$

we see that ϕ_1 cannot be a characteristic function of $K^{(2)}(x, s)$ and so $\phi_1 \neq \phi_2$ although λ_1 may equal λ_2. Continuing this procedure we shall find that either there exists an n such that

$$K^{(n+1)}(x, s) = K(x, s) - \sum_{\nu=1}^{n} \frac{\phi_\nu(x)\overline{\phi_\nu(s)}}{\lambda_\nu} \equiv 0$$

in which case we have the bilinear formula

$$K(x, s) = \sum_{\nu=1}^{n} \frac{\phi_\nu(x)\overline{\phi_\nu(s)}}{\lambda_\nu}. \tag{8}$$

or there exists an infinite number of characteristic values λ_ν and their associated characteristic functions ϕ_ν.

By Bessel's inequality (6.35) applied to $K(x, s)$ we have

$$\int_a^b |K(x, s)|^2 \, ds \geq \sum_{\nu=1}^{n} \left| \int_a^b K(x, s)\phi_\nu(s) \, ds \right|^2$$

$$= \sum_{\nu=1}^{n} \frac{|\phi_\nu(x)|^2}{\lambda_\nu^2} \tag{9}$$

so that, remembering that the ϕ_ν are normalized,

$$\|K\|_2^2 \geq \sum_{\nu=1}^{n} \frac{1}{\lambda_\nu^2}. \tag{10}$$

Hence if $|\lambda_\nu| < c$ for $\nu = 1, \ldots, n$ then $n \leq c^2 \|K\|_2^2$ and thus there are only a finite number of characteristic values in the interval $(-c, c)$. Consequently the spectrum of characteristic values is countable and has no finite point of accumulation.

We shall suppose that $|\lambda_1| \leq |\lambda_2| \leq \ldots$ and that if λ_ν is a characteristic value of rank p it will be repeated p times in this series and that the associated p linearly independent characteristic functions are orthogonal and normalized. We shall refer to these characteristic functions $\phi_1(x)$, $\phi_2(x)$, ... and their associated characteristic values $\lambda_1, \lambda_2, \ldots$ as a *full orthonormal system*. We note however that this system of functions need not be complete.

10.3 Expansion theorems

Although the series

$$S_n(x, s) = \sum_{\nu=1}^{n} \frac{\phi_\nu(x)\overline{\phi_\nu(s)}}{\lambda_\nu} \tag{11}$$

need not converge as $n \to \infty$, we can show that it is mean square convergent to $K(x, s)$, that is

$$\lim_{n \to \infty} \int_a^b \int_a^b |K(x, s) - S_n(x, s)|^2 \, dx \, ds = 0. \tag{12}$$

Thus, by the Riesz form of the Riesz-Fischer theorem stated in section 6.3.4, there exists a L^2 Hermitian kernel $Q(x, s)$ such that

$$\lim_{n \to \infty} \int_a^b |Q(x, s) - S_n(x, s)|^2 \, ds = 0 \qquad (a \le x \le b) \tag{13}$$

and

$$\int_a^b Q(x, s) \phi_\nu(s) \, ds = \frac{\phi_\nu(x)}{\lambda_\nu} \qquad (\nu = 1, 2, \ldots), \tag{14}$$

since

$$\sum_{\nu=1}^{\infty} \frac{|\phi_\nu(x)|^2}{\lambda_\nu^2} \le \int_a^b |K(x, s)|^2 \, ds < \infty \tag{15}$$

using (9) and that $K(x, s)$ is square integrable.

Hence, setting

$$P(x, s) = K(x, s) - Q(x, s) \tag{16}$$

we see that

$$\int_a^b P(x, s) \phi_\nu(s) \, ds = 0 \qquad (\nu = 1, 2, \ldots) \tag{17}$$

since $Q(x, s)$ has the same Fourier coefficients as $K(x, s)$.

In order to establish (12) we need to show that $P(x, s)$ is null. Let us suppose that, on the contrary, $P(x, s)$ is non-null. Then there exists a normalized characteristic function $\phi_0(x)$ with characteristic value λ_0 such that

$$\lambda_0 \boldsymbol{P} \phi_0 = \phi_0 \tag{18}$$

which gives

$$(\phi_0, \phi_\nu) = \lambda_0 (\boldsymbol{P} \phi_0, \phi_\nu) = \lambda_0 (\phi_0, \boldsymbol{P} \phi_\nu) = 0 \qquad (\nu = 1, 2, \ldots) \tag{19}$$

since $\boldsymbol{P}$ is Hermitian and using (17).

Now for any positive integer n we have

$$\int_a^b Q(x, s) \phi_0(s) \, ds = \int_a^b \left\{ Q(x, s) - S_n(x, s) \right\} \phi_0(s) \, ds \tag{20}$$

since $(\phi_0, \phi_\nu) = 0$ for $\nu = 1, 2, \ldots$ and so, given any $\varepsilon > 0$ there exists N such that for $n > N$

$$|\boldsymbol{Q}\phi_0|^2 \leqslant \int_a^b |Q(x, s) - S_n(x, s)|^2 \, \mathrm{d}s < \varepsilon \tag{21}$$

using Schwarz's inequality and (13). As $\boldsymbol{Q}\phi_0$ is independent of n it follows that $\boldsymbol{Q}\phi_0 = 0$ and so

$$\lambda_0 \boldsymbol{K}\phi_0 = \lambda_0 \boldsymbol{P}\phi_0 = \phi_0 \tag{22}$$

which means that ϕ_0 is a characteristic function of $\boldsymbol{K}$ but is orthogonal to all the ϕ_ν. This provides a contradiction since the ϕ_ν form a full orthonormal system. Hence $P(x, s)$ must be a null kernel and so

$$Q(x, s) = K(x, s) \tag{23}$$

which means that (13) holds with Q replaced by K, yielding the result (12) we wished to establish.

10.3.1 Hilbert-Schmidt theorem

If $f(x)$ is a square integrable function given by

$$f(x) = \int_a^b K(x, s) g(s) \, \mathrm{d}s \tag{24}$$

where $K(x, s)$ is a Hilbert-Schmidt kernel and $g(s)$ is a square integrable function, then the Hilbert-Schmidt theorem states that

$$f(x) = \sum_{\nu=1}^\infty a_\nu \phi_\nu(x) \tag{25}$$

where

$$a_\nu = (f, \phi_\nu) = \frac{1}{\lambda_\nu}(g, \phi_\nu) \tag{26}$$

and $\phi_\nu(x)$, λ_ν $(\nu = 1, 2, \ldots)$ are the characteristic functions and values of $K(x, s)$.

The *Hilbert-Schmidt series* (25) for $f(x)$ converges absolutely and uniformly if, in addition, the kernel satisfies

$$\int_a^b |K(x, s)|^2 \, \mathrm{d}s < A^2 \tag{27}$$

where A is a constant.

It is not difficult to establish (26) for we have

$$a_\nu = (f, \phi_\nu)$$

$$= \int_a^b \overline{\phi_\nu(x)} \, dx \int_a^b K(x, s) g(s) \, ds$$

$$= \int_a^b g(s) \, ds \int_a^b \overline{K(s, x)} \, \overline{\phi_\nu(x)} \, dx$$

$$= (g, \boldsymbol{K}\phi_\nu)$$

$$= \frac{1}{\lambda_\nu} (g, \phi_\nu).$$

Hence

$$f(x) = \int_a^b \{K(x, s) - S_n(x, s)\} g(s) \, ds + \sum_{\nu=1}^n \frac{\phi_\nu(x)}{\lambda_\nu} \int_a^b \overline{\phi_\nu(s)} \, g(s) \, ds$$

$$= \int_a^b \{K(x, s) - S_n(x, s)\} g(s) \, ds + \sum_{\nu=1}^n a_\nu \phi_\nu(x) \tag{28}$$

and so, employing Schwarz's inequality, we obtain

$$\left| f(x) - \sum_{\nu=1}^n a_\nu \phi_\nu(x) \right|^2 \le \int_a^b |K(x, s) - S_n(x, s)|^2 \, ds \int_a^b |g(s)|^2 \, ds. \tag{29}$$

Since the right-hand side of (29) can be made as small as we wish by letting n become sufficiently large, the result (25) is proved.

Setting

$$b_\nu = (g, \phi_\nu) \tag{30}$$

we have

$$\left\{ \sum_{\nu=n+1}^\infty |a_\nu \phi_\nu(x)| \right\}^2 = \left\{ \sum_{\nu=n+1}^\infty \left| \frac{b_\nu}{\lambda_\nu} \phi_\nu(x) \right| \right\}^2$$

$$\le \left\{ \sum_{\nu=n+1}^\infty |b_\nu|^2 \right\} \left\{ \sum_{\nu=n+1}^\infty \frac{|\phi_\nu(x)|^2}{\lambda_\nu^2} \right\} \tag{31}$$

using Cauchy's inequality (6.9). Now $g(x)$ is square integrable and so $\sum_{\nu=1}^\infty |b_\nu|^2 < \infty$. Hence, given any $\varepsilon > 0$ there exists N such that for $n > N$

$$\sum_{\nu=n+1}^\infty |b_\nu|^2 < \varepsilon^2. \tag{32}$$

Also (15) with the condition (27) gives

$$\sum_{\nu=n+1}^{\infty} \frac{|\phi_\nu(x)|^2}{\lambda_\nu^2} \leq \int_a^b |K(x, s)|^2 \, ds < A^2. \tag{33}$$

Thus for $n > N$ we have

$$\sum_{\nu=n+1}^{\infty} |a_\nu \phi_\nu(x)| < \varepsilon A \tag{34}$$

from which it follows that the Hilbert-Schmidt series (25) is absolutely and uniformly convergent.

10.3.2 Hilbert's formula

As a corollary to the Hilbert-Schmidt theorem proved above, we have that if $g(x)$ and $h(x)$ are both square integrable functions then

$$(\boldsymbol{K}g, h) = \sum_{\nu=1}^{\infty} \frac{1}{\lambda_\nu} (g, \phi_\nu)(\phi_\nu, h) \tag{35}$$

which is Hilbert's formula. It follows from the Hilbert-Schmidt series (25) for $f = \boldsymbol{K}g$ by taking the inner product with h. This can be done term by term on the right-hand side since the series is uniformly convergent.

For $h = g$ Hilbert's formula reduces to

$$(\boldsymbol{K}g, g) = \sum_{\nu=1}^{\infty} \frac{1}{\lambda_\nu} |(g, \phi_\nu)|^2. \tag{36}$$

10.3.3 Expansion theorem for iterated kernels

Since the iterated kernel $K_n(x, s)$ for $n \geq 2$ is given by

$$K_n(x, s) = \int_a^b K(x, t) K_{n-1}(t, s) \, dt$$

and so is of the form (24) with $g = K_{n-1}$, it follows from the Hilbert-Schmidt theorem that if $K(x, s)$ is a Hilbert-Schmidt kernel then

$$K_n(x, s) = \sum_{\nu=1}^{\infty} a_\nu(s) \phi_\nu(x)$$

where

$$a_\nu(s) = \int_a^b K_n(x, s) \overline{\phi_\nu(x)} \, dx = \frac{1}{\lambda_\nu^n} \overline{\phi_\nu(s)}.$$

Hence

$$K_n(x, s) = \sum_{\nu=1}^{\infty} \frac{\phi_\nu(x)\overline{\phi_\nu(s)}}{\lambda_\nu^n} \quad (n \geq 2) \tag{37}$$

where the series is absolutely and uniformly convergent if the kernel $K(x, s)$ satisfies the condition (27).

We deduce that

$$\kappa_n = \text{trace } (\boldsymbol{K}^n) = \int_a^b K_n(x, x)\, \mathrm{d}x$$

$$= \sum_{\nu=1}^{\infty} \frac{1}{\lambda_\nu^n} \quad (n \geq 2) \tag{38}$$

and in particular we have

$$\kappa_2 = \|\boldsymbol{K}\|_2^2 = \sum_{\nu=1}^{\infty} \frac{1}{\lambda_\nu^2} < \infty. \tag{39}$$

10.4 Solution of Fredholm equation of second kind

Consider the Fredholm equation of the second kind

$$\phi(x) - f(x) = \lambda \int_a^b K(x, s)\phi(s)\, \mathrm{d}s \tag{40}$$

where $K(x, s)$ is a Hilbert-Schmidt kernel satisfying the condition (27), the function $f(x)$ is square integrable and λ is a regular value.

If the solution $\phi(x)$ of (40) is square integrable then the Hilbert-Schmidt theorem gives

$$\phi(x) - f(x) = \sum_{\nu=1}^{\infty} a_\nu \phi_\nu(x)$$

where

$$a_\nu = (\phi - f, \phi_\nu) = (\phi, \phi_\nu) - (f, \phi_\nu)$$

and also

$$a_\nu = \lambda(\boldsymbol{K}\phi, \phi_\nu) = \frac{\lambda}{\lambda_\nu}(\phi, \phi_\nu)$$

Hence

$$(\phi, \phi_\nu) = \frac{\lambda_\nu}{\lambda_\nu - \lambda}(f, \phi_\nu)$$

and

$$a_\nu = \frac{\lambda}{\lambda_\nu - \lambda}(f, \phi_\nu).$$

Consequently the solution can be expressed as the absolutely and uniformly convergent series

$$\phi(x) = f(x) + \lambda \sum_{\nu=1}^{\infty} \frac{(f, \phi_\nu)\phi_\nu(x)}{\lambda_\nu - \lambda}$$

$$= f(x) + \lambda \sum_{\nu=1}^{\infty} \int_a^b \frac{\phi_\nu(x)\overline{\phi_\nu(s)}}{\lambda_\nu - \lambda} f(s) \, ds \qquad (41)$$

and so the resolvent kernel takes the form

$$R(x, s; \lambda) = \sum_{\nu=1}^{\infty} \frac{\phi_\nu(x)\overline{\phi_\nu(s)}}{\lambda_\nu - \lambda} \qquad (42)$$

if this series is uniformly convergent, for then it is permissible to reverse the order of the integration and the summation in (41).

Now we have shown in section 10.3 that $\sum_{\nu=1}^{\infty} |\phi_\nu(x)|^2/\lambda_\nu^2$ is convergent, and since $\lambda_\nu/(\lambda_\nu - \lambda) \to 1$ as $\nu \to \infty$ it follows that

$$\sum_{\nu=1}^{\infty} \frac{|\phi_\nu(x)|^2}{(\lambda_\nu - \lambda)^2} < \infty. \qquad (43)$$

Hence, by the Riesz-Fischer theorem, the series (42) is mean square convergent to a square integrable kernel $R(x, s; \lambda)$. Then using the Hilbert-Schmidt theorem and the resolvent equation (8.12) written in the form

$$R(x, s; \lambda) = K(x, s) + \lambda \int_a^b K(x, t)R(t, s; \lambda) \, dt, \qquad (44)$$

we obtain the absolutely and uniformly convergent series for the resolvent kernel

$$R(x, s; \lambda) = K(x, s) + \lambda \sum_{\nu=1}^{\infty} \frac{\phi_\nu(x)\overline{\phi_\nu(s)}}{\lambda_\nu(\lambda_\nu - \lambda)} \qquad (45)$$

which enables us to express the solution of (40) as

$$\phi(x) = f(x) + \lambda \int_a^b K(x, s)f(s) \, ds + \lambda^2 \sum_{\nu=1}^{\infty} \frac{(f, \phi_\nu)\phi_\nu(x)}{\lambda_\nu(\lambda_\nu - \lambda)}. \qquad (46)$$

It can be seen from (45) that the singularities of the resolvent kernel are simple poles occurring at the characteristic values λ_ν of the Hermitian kernel $K(x, s)$.

10.5   Bounds on characteristic values

Using Hilbert's formula (36) for a square integrable function $\psi(x)$ we have

$$(\boldsymbol{K}\psi, \psi) = \sum_{\nu=1}^{\infty} \frac{|(\psi, \phi_\nu)|^2}{\lambda_\nu} \tag{47}$$

and clearly

$$(\boldsymbol{K}\phi_\nu, \phi_\nu) = \frac{1}{\lambda_\nu} \qquad (\nu = 1, 2, \ldots) \tag{48}$$

Let λ_ν^+, λ_ν^- ($\nu = 1, 2, \ldots$) denote the positive and negative characteristic values associated with the characteristic functions $\phi_\nu^+(x)$, $\phi_\nu^-(x)$ respectively, and arranged so that

$$\begin{aligned} 0 &< \lambda_1^+ \leq \lambda_2^+ \leq \ldots \leq \lambda_\nu^+ \leq \ldots, \\ 0 &> \lambda_1^- \geq \lambda_2^- \geq \ldots \geq \lambda_\nu^- \geq \ldots. \end{aligned} \tag{49}$$

Hence

$$(\boldsymbol{K}\psi, \psi) \leq \sum_{\nu=1}^{\infty} \frac{|(\psi, \phi_\nu^+)|^2}{\lambda_\nu^+} \leq \frac{1}{\lambda_1^+} \sum_{\nu=1}^{\infty} |(\psi, \phi_\nu^+)|^2$$

$$\leq \frac{1}{\lambda_1^+} (\psi, \psi) \tag{50}$$

and so, introducing the functional

$$I[\psi] = \frac{(\boldsymbol{K}\psi, \psi)}{(\psi, \psi)} \tag{51}$$

we obtain

$$I[\psi] \leq \frac{1}{\lambda_1^+}, \tag{52}$$

the equality being attained for $\psi = \phi_1^+$.

Now suppose that $\psi(x)$ is orthogonal to the characteristic functions $\phi_1^+(x), \phi_2^+(x), \ldots, \phi_{n-1}^+(x)$ so that

$$(\psi, \phi_1^+) = (\psi, \phi_2^+) = \ldots = (\psi, \phi_{n-1}^+) = 0.$$

Then

$$(\boldsymbol{K}\psi, \psi) \le \sum_{\nu=n}^{\infty} \frac{|(\psi, \phi_\nu^+)|^2}{\lambda_\nu^+} \le \frac{1}{\lambda_n^+} \sum_{\nu=n}^{\infty} |(\psi, \phi_\nu^+)|^2$$

$$\le \frac{1}{\lambda_n^+} (\psi, \psi) \tag{53}$$

which gives

$$I[\psi] \le \frac{1}{\lambda_n^+}. \tag{54}$$

Similarly we can show that

$$I[\psi] \ge \frac{1}{\lambda_1^-} \tag{55}$$

and that if $\psi(x)$ is orthogonal to $\phi_1^-, \phi_2^-, \ldots, \phi_{n-1}^-$ then

$$I[\psi] \ge \frac{1}{\lambda_n^-}. \tag{56}$$

10.6   Positive kernels

A Hermitian kernel $K(x, s)$ is said to be *positive* or *non-negative definite* if

$$(\boldsymbol{K}\psi, \psi) \ge 0 \tag{57}$$

for every square integrable function $\psi(x)$; and said to be *positive definite* if

$$(\boldsymbol{K}\psi, \psi) > 0 \tag{58}$$

for every square integrable function satisfying $\|\psi\| > 0$.

We see at once from (47) and (48) that $K(x, s)$ is a positive kernel if and only if all its characteristic values are positive, that is $0 < \lambda_1 \le \lambda_2 \le \ldots$.

Further $K(x, s)$ is positive definite if and only if the full orthonormal system of characteristic functions $\phi_\nu(x)$ $(\nu = 1, 2, \ldots)$ is also complete. For $(\boldsymbol{K}\psi, \psi) = 0$ if and only if $(\psi, \phi_\nu) = 0$ for all ν.

Now let us suppose that the positive kernel $K(x, s)$ is continuous. Since $K(x, s)$ is Hermitian it follows that $K(x, x)$ is real. Then we can show that $K(x, x) \ge 0$ for $a \le x \le b$.

To this end we suppose that, on the contrary, there exists x_0 with $a < x_0 < b$ such that $K(x_0, x_0) = -\varepsilon$ where $\varepsilon > 0$. We are given that

$K(x, s)$ is continuous. Hence the real part of the kernel must also be continuous and so we can find $\delta > 0$ such that for $x_0 - \delta \leqslant x \leqslant x_0 + \delta$, $x_0 - \delta \leqslant s \leqslant x_0 + \delta$ we have $\text{Re}\,\{K(x, s)\} < -\tfrac{1}{2}\varepsilon$.

Let

$$\psi_0(x) = \begin{cases} 1 & (x_0 - \delta \leqslant x \leqslant x_0 + \delta) \\ 0 & \text{elsewhere} \end{cases}$$

and then, since $(\boldsymbol{K}\psi_0, \psi_0)$ is real, we see that

$$\begin{aligned}
(\boldsymbol{K}\psi_0, \psi_0) &= \int_a^b \int_a^b \text{Re}\,\{K(x, s)\}\psi_0(s)\psi_0(x)\,\mathrm{d}s\,\mathrm{d}x \\
&= \int_{x_0-\delta}^{x_0+\delta} \int_{x_0-\delta}^{x_0+\delta} \text{Re}\,\{K(x, s)\}\,\mathrm{d}s\,\mathrm{d}x \\
&< -\tfrac{1}{2}\varepsilon(2\delta)^2 < 0
\end{aligned}$$

which contradicts our supposition that the kernel $K(x, s)$ is positive. Hence $K(x, x) \geqslant 0$ for $a < x < b$ and also at the end points since the kernel is continuous.

10.7   Mercer's theorem

Mercer showed in 1909 that the expansion theorem given in section 10.3 can be strengthened if $K(x, s)$ is a continuous positive kernel. Thus, supposing that $K(x, s)$ is such a kernel, we see that

$$K^{(n+1)}(x, s) = K(x, s) - \sum_{\nu=1}^n \frac{\phi_\nu(x)\overline{\phi_\nu(s)}}{\lambda_\nu} \tag{59}$$

is also positive, and continuous for all n since the characteristic functions $\phi_\nu(x)$ of a continuous kernel are continuous. Hence, using the result proved in the previous section, we have

$$K(x, x) - \sum_{\nu=1}^n \frac{\phi_\nu(x)\overline{\phi_\nu(x)}}{\lambda_\nu} = K^{(n+1)}(x, x) \geqslant 0 \tag{60}$$

giving

$$\sum_{\nu=1}^n \frac{|\phi_\nu(x)|^2}{\lambda_\nu} \leqslant K(x, x) \tag{61}$$

for all n. Integrating we obtain

$$\sum_{\nu=1}^n \frac{1}{\lambda_\nu} \leqslant \int_a^b K(x, x)\,\mathrm{d}x = \text{trace } K = \kappa_1 \tag{62}$$

and so, as $K(x, x)$ is continuous and therefore bounded, it follows that

$$\sum_{\nu=1}^{\infty} \frac{1}{\lambda_\nu} < \infty \qquad (63)$$

which is a much stronger result than (39).

Now, by Cauchy's inequality (6.9), we have

$$\left\{ \sum_{\nu=n+1}^{m} \frac{|\phi_\nu(x)\overline{\phi_\nu(s)}|}{\lambda_\nu} \right\}^2 \leq \left\{ \sum_{\nu=n+1}^{m} \frac{|\phi_\nu(x)|^2}{\lambda_\nu} \right\} \left\{ \sum_{\nu=n+1}^{m} \frac{|\phi_\nu(s)|^2}{\lambda_\nu} \right\}. \qquad (64)$$

Also (61) implies that $\sum_{\nu=1}^{\infty} \frac{|\phi_\nu(x)|^2}{\lambda_\nu}$ is convergent for all x. Since $K(x, x)$ is bounded

$$\sum_{\nu=n+1}^{m} \frac{|\phi_\nu(x)|^2}{\lambda_\nu} < C^2 \qquad (a \leq x \leq b) \qquad (65)$$

where C is a positive constant, and given any $\varepsilon > 0$ and a fixed value of s there exists N such that

$$\sum_{\nu=n+1}^{m} \frac{|\phi_\nu(s)|^2}{\lambda_\nu} < \varepsilon^2 \qquad (66)$$

for $n, m > N$. Thus we have

$$\sum_{\nu=n+1}^{m} \frac{|\phi_\nu(x)\overline{\phi_\nu(s)}|}{\lambda_\nu} < \varepsilon C \qquad (67)$$

and so we see that

$$\sum_{\nu=1}^{\infty} \frac{\phi_\nu(x)\overline{\phi_\nu(s)}}{\lambda_\nu} \qquad (68)$$

is absolutely and uniformly convergent in x for each s. Similarly we can show that (68) is absolutely and uniformly convergent in s for each x.

Since we know that (68) is mean square convergent to $K(x, s)$ it follows that we must have

$$K(x, s) = \sum_{\nu=1}^{\infty} \frac{\phi_\nu(x)\overline{\phi_\nu(s)}}{\lambda_\nu} \qquad (69)$$

In particular

$$K(x, x) = \sum_{\nu=1}^{\infty} \frac{|\phi_\nu(x)|^2}{\lambda_\nu} \qquad (70)$$

where the partial sums of the infinite series form a monotonic sequence of continuous functions which is convergent to a continuous function.

Now Dini's theorem states that if a monotonic sequence of real continuous functions is convergent in the closed interval $a \leq x \leq b$ to a continuous function, then the sequence converges uniformly in this interval.

It follows that the series (70) is uniformly convergent in x and hence, by Cauchy's inequality, that (69) is uniformly convergent in x, s.

10.8   Variational principles

We showed in section 10.5 that if $\boldsymbol{K}$ is an integral operator possessing a square integrable Hermitian kernel and ψ is a square integrable function, then the functional

$$I[\psi] = \frac{(\boldsymbol{K}\psi, \psi)}{(\psi, \psi)} \tag{71}$$

has the maximum value $1/\lambda_1^+$ when $\psi = \phi_1^+$ and the minimum value $1/\lambda_1^-$ when $\psi = \phi_1^-$; and moreover if ψ is orthogonal to the characteristic functions $\phi_1^+, \phi_2^+, \ldots, \phi_{n-1}^+$ then $I[\psi]$ has the maximum value $1/\lambda_n^+$ when $\psi = \phi_n^+$, while if ψ is orthogonal to $\phi_1^-, \phi_2^-, \ldots, \phi_{n-1}^-$ then $I[\psi]$ has the minimum value $1/\lambda_n^-$ when $\psi = \phi_n^-$. These properties lead us to expect that $I[\psi]$ is stationary whenever ψ is one of the characteristic functions ϕ_ν of $\boldsymbol{K}$, thus giving rise to a variational principle.

To establish that this is the case we examine the variation

$$\delta I[\phi_\nu] = I[\phi_\nu + \delta\phi_\nu] - I[\phi_\nu] \tag{72}$$

arising from an arbitrary variation $\delta\phi_\nu$ in the characteristic function ϕ_ν. We have

$$
\begin{aligned}
I[\phi_\nu + \delta\phi_\nu] &= \frac{(\boldsymbol{K}\phi_\nu + \boldsymbol{K}\,\delta\phi_\nu,\ \phi_\nu + \delta\phi_\nu)}{(\phi_\nu + \delta\phi_\nu,\ \phi_\nu + \delta\phi_\nu)} \\
&= I[\phi_\nu] + \frac{(\boldsymbol{K}\phi_\nu,\ \delta\phi_\nu) + (\boldsymbol{K}\,\delta\phi_\nu,\ \phi_\nu)}{(\phi_\nu,\ \phi_\nu)} \\
&\quad - \frac{(\boldsymbol{K}\phi_\nu,\ \phi_\nu)\{(\phi_\nu,\ \delta\phi_\nu) + (\delta\phi_\nu,\ \phi_\nu)\}}{(\phi_\nu,\ \phi_\nu)^2} + O\{|\delta\phi_\nu|^2\} \\
&= I[\phi_\nu] + O\{|\delta\phi_\nu|^2\}
\end{aligned}
\tag{73}
$$

since $\lambda_\nu K \phi_\nu = \phi_\nu$. Neglecting the quantity $O\{|\delta\phi_\nu|^2\}$ of the second order of smallness, we may write this result as the variational principle

$$\delta I[\phi_\nu] = 0 \tag{74}$$

which establishes that $I[\psi]$ is stationary when $\psi = \phi_\nu$.

We can also derive a variational principle for (f, ϕ) where ϕ is the solution of the inhomogeneous equation

$$\phi = f + \lambda K \phi \tag{75}$$

choosing λ to be real so that

$$(f, \phi) = (\phi, \phi) - \lambda(K\phi, \phi) = (\phi, f) \tag{76}$$

and thus is a real quantity also.

We introduce the functional

$$J[\psi] = (f, \psi) + (\psi, f) - (\psi, \psi) + \lambda(K\psi, \psi) \tag{77}$$

and so we have $J[\phi] = (f, \phi)$. Then

$$\begin{aligned}
J[\phi + \delta\phi] &= (f, \phi + \delta\phi) + (\phi + \delta\phi, f) - (\phi + \delta\phi, \phi + \delta\phi) \\
&\quad + \lambda(K\phi + K\delta\phi, \phi + \delta\phi) \\
&= J[\phi] + (f - \phi + \lambda K\phi, \delta\phi) \\
&\quad + (\delta\phi, f - \phi + \lambda K\phi) - (\{1 - \lambda K\}\delta\phi, \delta\phi) \\
&= J[\phi] - (\{1 - \lambda K\}\delta\phi, \delta\phi)
\end{aligned} \tag{78}$$

since ϕ satisfies (75). Neglecting the second order quantity on the right-hand side of (78) yields the variational principle

$$\delta J[\phi] = 0. \tag{79}$$

Moreover if λK is a negative operator we see that

$$\delta J[\phi] \leq 0 \tag{80}$$

and this means that $J[\psi]$ attains a maximum value for $\psi = \phi$ so that $J[\psi]$ provides a lower bound to (f, ϕ).

An alternative expression for $J[\psi]$ may be obtained by setting $\psi = \alpha\chi$ and optimizing with respect to the parameter α. We have

$$J[\alpha\chi] = \bar{\alpha}(f, \chi) + \alpha(\chi, f) - \alpha\bar{\alpha}(\chi, \chi) + \lambda\alpha\bar{\alpha}(K\chi, \chi),$$

and taking

$$\frac{\partial J}{\partial \alpha} = \frac{\partial J}{\partial \bar{\alpha}} = 0$$

we find that

$$\alpha = \frac{(f, \chi)}{(\chi, \chi) - \lambda(\boldsymbol{K}\chi, \chi)}.$$

This results in the homogeneous formula

$$J[\chi] = \frac{(f, \chi)(\chi, f)}{(\chi, \chi) - \lambda(\boldsymbol{K}\chi, \chi)} \tag{81}$$

and clearly $J[\phi] = (f, \phi)$ using (76).

10.8.1 Rayleigh-Ritz variational method

We may use the variational principle (74) to devise a method of obtaining approximations to the characteristic values. This method is associated with the original work of Rayleigh and Ritz in connection with the determination of the natural frequencies of vibration of mechanical systems.

Let $\psi_1, \psi_2, \ldots, \psi_n$ be a set of n linearly independent square integrable functions and put

$$\psi = \sum_{r=1}^{n} c_r \psi_r \tag{82}$$

where the c_r $(r = 1, \ldots, n)$ are n adjustable parameters. Then we have

$$I[\psi] = \frac{\sum_{r=1}^{n} \sum_{s=1}^{n} c_r \bar{c}_s k_{rs}}{\sum_{r=1}^{n} \sum_{s=1}^{n} c_r \bar{c}_s a_{rs}} \tag{83}$$

where $k_{rs} = (\boldsymbol{K}\psi_r, \psi_s)$ and $a_{rs} = (\psi_r, \psi_s)$. We now optimize with respect to the parameters c_r by taking

$$\frac{\partial I}{\partial c_r} = \frac{\partial I}{\partial \bar{c}_r} = 0 \qquad (r = 1, \ldots, n) \tag{84}$$

and these yield the set of n linear homogeneous equations

$$\lambda \sum_{s=1}^{n} k_{rs} \bar{c}_s - \sum_{s=1}^{n} a_{rs} \bar{c}_s = 0 \qquad (r = 1, \ldots, n) \tag{85}$$

where $1/\lambda$ is the value of $I[\psi]$. These equations have a non-trivial solution if and only if the determinant of the coefficients vanishes:

$$
\begin{vmatrix}
\lambda k_{11} - a_{11} & \lambda k_{12} - a_{12} & \cdots & \lambda k_{1n} - a_{1n} \\
\lambda k_{21} - a_{21} & \lambda k_{22} - a_{22} & \cdots & \lambda k_{2n} - a_{2n} \\
\cdot & \cdot & & \cdot \\
\cdot & \cdot & & \cdot \\
\cdot & \cdot & & \cdot \\
\lambda k_{n1} - a_{n1} & \lambda k_{n2} - a_{n2} & \cdots & \lambda k_{nn} - a_{nn}
\end{vmatrix} = 0 \qquad (86)
$$

The n roots $\lambda^{(1)}, \lambda^{(2)}, \ldots, \lambda^{(n)}$ of this secular equation provide variational approximations to the characteristic values of K. In particular if K is positive and $\lambda^{(1)}$ is the least of these roots then $0 < \lambda_1 \leqslant \lambda^{(1)}$, thus giving an upper bound to the smallest characteristic value.

Example. As a simple illustration of the Rayleigh-Ritz method we consider a vibrating flexible string having uniform density ρ and length l which is stretched at tension T and fixed at its two ends $x = 0, l$.

Referring to section 2.1 we see that the homogeneous Fredholm linear integral equation corresponding to this physical problem has the positive symmetric kernel

$$
K(x, s) = \begin{cases} \dfrac{s(l-x)}{l} & (0 \leqslant s \leqslant x \leqslant l) \\[2mm] \dfrac{x(l-s)}{l} & (0 \leqslant x \leqslant s \leqslant l) \end{cases} \qquad (87)
$$

Let us take

$$
\psi(x) = c_1 x + c_2 x^2 \qquad (88)
$$

as our *trial function*. Then

$$
a_{11} = \frac{l^3}{3}, \qquad a_{12} = a_{21} = \frac{l^4}{4}, \qquad a_{22} = \frac{l^5}{5} \qquad (89)
$$

and

$$
k_{11} = \frac{l^5}{45}, \qquad k_{12} = k_{21} = \frac{l^6}{72}, \qquad k_{22} = \frac{l^7}{112} \qquad (90)
$$

The secular equation is

$$\begin{vmatrix} \dfrac{\lambda l^5}{45} - \dfrac{l^3}{3} & \dfrac{\lambda l^6}{72} - \dfrac{l^4}{4} \\[2ex] \dfrac{\lambda l^6}{72} - \dfrac{l^4}{4} & \dfrac{\lambda l^7}{112} - \dfrac{l^5}{5} \end{vmatrix} = 0 \tag{91}$$

which gives the quadratic equation

$$5(\lambda l^2)^2 - 432\lambda l^2 + 3780 = 0 \tag{92}$$

whose least root is $\lambda^{(1)} = 9{\cdot}880/l^2$. Then the approximate least angular frequency of vibration resulting from the application of the Rayleigh-Ritz method to the function (88) is given by

$$\{\omega^{(1)}\}^2 = \frac{T}{\rho}\lambda^{(1)} = 9{\cdot}880\frac{T}{\rho l^2} \tag{93}$$

which is only slightly greater than the exact value

$$\omega_1^2 = \pi^2 \frac{T}{\rho l^2} = 9{\cdot}870\frac{T}{\rho l^2}. \tag{94}$$

Problems

1. Show that the characteristic values of a skew-Hermitian kernel (called skew-symmetric if real) satisfying

$$K(x, s) = -\overline{K(s, x)}$$

are pure imaginary numbers.

2. Show that the Poisson kernel

$$K(x, s) = \frac{1}{2\pi}\frac{1-a^2}{1+a^2-2a\cos(x-s)} \qquad (0 \le x \le 2\pi, 0 \le s \le 2\pi)$$

where $|a| < 1$, has the characteristic functions

$$\frac{1}{\sqrt{2\pi}}, \qquad \frac{1}{\sqrt{\pi}}\sin \nu x, \qquad \frac{1}{\sqrt{\pi}}\cos \nu x \qquad (\nu = 1, 2, \dots)$$

with characteristic values 1, $a^{-\nu}$, $a^{-\nu}$ respectively.

Use the expansion theorem (42) for the resolvent kernel to show that it is given by the series obtained in problem 4 at the end of chapter 8.

3. Show that the homogeneous equation

$$\phi(x) = \lambda \int_0^1 K(x, s)\phi(s) \, ds$$

with the symmetric kernel

$$K(x, s) = \begin{cases} x(1-s) & (x \leq s) \\ s(1-x) & (x \geq s) \end{cases}$$

is equivalent to the differential equation

$$\phi'' + \lambda\phi = 0$$

with $\phi(0) = \phi(1) = 0$.

Hence show that the kernel has characteristic functions $\sqrt{2} \sin \nu\pi x \ (\nu = 1, 2, \ldots)$ with characteristic values $(\nu\pi)^2$ respectively. Use the expansion theorem (69) to show that the kernel $K(x, s)$ is given by the uniformly convergent series

$$\frac{2}{\pi^2} \sum_{\nu=1}^{\infty} \frac{\sin \nu\pi x \sin \nu\pi s}{\nu^2}$$

and also obtain an expansion for the resolvent kernel.

4. Show that the homogeneous equation with the symmetric kernel

$$K(x, s) = \begin{cases} x & (0 \leq x \leq s \leq 1) \\ s & (0 \leq s \leq x \leq 1) \end{cases}$$

is equivalent to the differential equation

$$\phi'' + \lambda\phi = 0$$

with $\phi(0) = \phi'(1) = 0$.

Hence show that the kernel has characteristic functions $\sqrt{2} \sin \{[(2\nu-1)/2]\pi x\} \ (\nu = 1, 2, \ldots)$ with characteristic values $[(2\nu-1)/2]^2 \pi^2$. Use the expansion theorems to obtain uniformly convergent series for $K(x, s)$ and the resolvent kernel.

5. Use the Rayleigh-Ritz variational method together with the trial function

$$\psi(x) = c_1 + c_2 x$$

to obtain an upper bound to the characteristic value for the separable kernel

$$K(x, s) = e^{x+s} \qquad (0 \leq x \leq 1, 0 \leq s \leq 1)$$

Verify that the result is slightly greater than the exact value obtained in problem 3 at the end of chapter 1.

6. Use the Rayleigh-Ritz variational method and the trial function

$$\psi(x) = c_1 x + c_2 x^2$$

to obtain an upper bound to the least characteristic value for the symmetric kernel in problem 4 and verify that it is slightly greater than the exact value $(\pi/2)^2$.

7. By using the variational principle (79) together with the trial function

$$\psi(x) = f(x) + \sum_{\nu=1}^{\infty} c_\nu \phi_\nu(x),$$

where the $\phi_\nu(x)$ are the characteristic functions of the square integrable Hermitian kernel $K(x, s)$ associated with the characteristic values λ_ν, show that

$$c_\nu = \frac{\lambda(f, \phi_\nu)}{\lambda_\nu - \lambda}$$

in agreement with the solution of the Fredholm equation of the second kind obtained in section 10.4.

Show also that for real λ

$$(f, \phi) = (f, f) + \lambda \sum_{\nu=1}^{\infty} \frac{(f, \phi_\nu)(\phi_\nu, f)}{\lambda_\nu - \lambda} = (\phi, f).$$

Bibliography

Bôcher, M., *An Introduction to the Study of Integral Equations*, Cambridge Tracts in Mathematics and Mathematical Physics, No. 10, Cambridge University Press, 1913.

Cochran, J. A., *The Analysis of Linear Integral Equations*, McGraw-Hill, New York, 1972.

Courant, R. and Hilbert, D., *Methods of Mathematical Physics*, Vol. 1, Ch. III, Interscience, New York, 1953.

Goursat, E., *A Course in Mathematical Analysis*, Vol. III, Part 2, Chs. VIII–XI, Dover, New York, 1964.

Green, C. D., *Integral Equation Methods*, Nelson, London, 1969.

Hildebrand, F. B., *Methods of Applied Mathematics*, 2nd edn., Ch. 3, Prentice-Hall, Englewood Cliffs, New Jersey, 1965.

Hochstadt, H., *Integral Equations*, Wiley, New York, 1973.

Hoheisel, G., *Integral Equations*, Nelson, London, 1967.

Kanwal, R.P., *Linear Integral Equations*, Academic Press, New York, 1971.

Lovitt, W. V., *Linear Integral Equations*, Dover, New York, 1950.

Mikhlin, S. G., *Integral Equations*, 2nd edn., Pergamon Press, London, 1964.

Morse, P. M. and Feshbach, H., *Methods of Theoretical Physics*, McGraw-Hill, New York, 1953.

Schmeidler, W., *Linear Operators in Hilbert Space*, Academic Press, New York, 1965.

Smirnov, V. I., *A Course of Higher Mathematics*, Vol. IV, *Integral Equations and Partial Differential Equations*, Pergamon Press, London, 1964.

Smithies, F., *Integral Equations*, Cambridge Tracts in Mathematics and Mathematical Physics, No. 49, Cambridge University Press, London, 1958.

Titchmarsh, E. C., *Introduction to the Theory of Fourier Integrals*, 2nd edn., Ch. 11, Oxford University Press, 1948.

Tricomi, F. G., *Integral Equations*, Interscience, New York, 1957.

Whittaker, E. T. and Watson, G. N., *A Course of Modern Analysis*, 4th edn., Ch. XI, Cambridge University Press, London, 1927.

Zabreyko, P. P., Koshelev, A. I., Krasnosel'skii, M. A., Mikhlin, S. G., Rakovshchik, L. S. and Stet'senko, V. Ya., *Integral Equations – A Reference Text*, Noordhoff International Publishing, Leyden, 1975.

Index

A LONGMAN MATHEMATICAL

Integral equations

This is a text for final year honours undergraduates of mathematics and mathematical physics.

Many mathematical problems, particularly in applied mathematics, can be formulated in two distinct but related ways, namely as differential equations or as integral equations. In the integral equation approach the boundary conditions are included specifically and this confers a valuable advantage to the method. Moreover the integral equation approach leads naturally to the solution of the problem, under suitable conditions, in the form of an infinite series. Integral equations are of considerable importance in the history of mathematics. Thus Laplace and Fourier transforms provide well known examples of the first kind. This book is mainly concerned with linear integral equations. It begins with a straightforward account of the subject, accompanied by simple examples of different types of integral equations and the methods of their solution, but goes on to provide a rather more abstract treatment with a discussion of linear operators in Hilbert space.

The author

Benjamin Lawrence Moiseiwitsch, B.Sc., Ph.D. (London), is Professor of Applied Mathematics at The Queen's University of Belfast. He is a Member of the Royal Irish Academy.

The editors

The editors of the *Longman Mathematical Texts* series are Professor Alan Jeffrey of the University of Newcastle upon Tyne and Dr Iain Adamson of the University of Dundee.

ISBN 0 582 44288 5